MZ ETZ Models Owners Workshop Manual

by Mark Coombs
with an additional Chapter on the 1991-on models
by Penny Cox

Models covered
ETZ125 Standard. 123cc. March 1986 to January 1992
ETZ125 Luxus. 123cc. March 1986 to September 1990,
   January 1992 on
ETZ125XR. 123cc. January 1988 to August 1989
ETZ125 Saxon Roadstar and Sportstar. 123cc. January 1993 on
ETZ150 Luxus. 143cc. March 1986 to October 1991
ETZ150 Saxon Roadstar. 143cc. June 1994 on
ETZ250 Standard. 243cc. October 1981 to August 1983
ETZ250 Luxus. 243cc. October 1981 to December 1990
ETZ250XR. 243cc. January 1988 to August 1989
ETZ251 Luxus. 243cc. July 1989 on
ETZ251 Econo. 243cc. July 1992 to January 1993
ETZ251 Saxon Tour and Fun. 243cc. January 1993 on
ETZ300 Luxus. 296cc. July 1987 to April 1990
ETZ301 Luxus. 291cc. December 1991 to January 1995
ETZ301 Saxon Tour and Fun. 291cc. January 1993 on

(1680-3U1)

Haynes Group Limited
Haynes North America, Inc

www.haynes.com

Acknowledgements

Our thanks are due to APS Motorcycles of Wells, MD Motorcycles of Wellington, Ken Doyle Motorcycles of Templecombe, and Kellaway Motorcycles of Bristol who provided both technical assistance and the machines featured in this manual. Thanks are also due to MZ GB Ltd for their assistance, and VEB Motorradwerk, Zschopau, for permission to reproduce some of their illustrations.

The Avon Rubber Company supplied information on tyre care and fitting, and NGK Spark Plugs (UK) Ltd provided information on plug maintenance and electrode conditions.

A book in the **Haynes Owners Workshop Manual Series**

ISBN 978 1 85960 065 8

British Library Cataloguing in Publication Data
A catalogue record for this book is available from the British Library

Contents

The 1990 MZ ETZ125 model

The 1990 MZ ETZ251 model

About this manual

The purpose of this manual is to present the owner with a concise and graphic guide which will enable him to tackle any operation from basic routine maintenance to a major overhaul. It has been assumed that any work would be undertaken without the luxury of a well-equipped workshop and a range of manufacturer's service tools.

To this end, the machine featured in the manual was stripped and rebuilt in our own workshop, by a team comprising a mechanic, a photographer and the author. The resulting photographic sequence depicts events as they took place, the hands shown being those of the mechanic and the author.

The use of specialised, and expensive, service tools was avoided unless their use was considered to be essential due to risk of breakage or injury. There is usually some way of improvising a method of removing a stubborn component, providing that a suitable degree of care is exercised.

The author learnt his motorcycle mechanics over a number of years, faced with the same difficulties and using similar facilities to those encountered by most owners. It is hoped that this practical experience can be passed on through the pages of this manual. Armed with a working knowledge of the machine, the author undertakes a consider-able amount of research in order that the maximum amount of data can be included in the manual.

A comprehensive section, preceding the main part of the manual, describes procedures for carrying out the routine maintenance of the machine at intervals of time and mileage. This section is included particularly for those owners who wish to ensure the efficient day-to-day running of their motorcycle, but who choose not to undertake overhaul or renovation work.

Each chapter is divided into numbered sections. Within these sections are numbered paragraphs. All photographs in the main chapters are captioned with a section/paragraph number to which they refer.

Figures (line illustrations) appear in a numerical but logical order, within a given chapter. Fig. 1.1 therefore refers to the first figure in chapter 1.

Left-hand and right-hand descriptions of the machine and its components refer to the left and right of a given machine when the rider is seated in the normal riding position.

Motorcycle manufacturers continually make changes to specifications and recommendations, and these, when notified, are incorporated into our manuals at the earliest opportunity.

Introduction to the MZ ETZ models

The history of VEB Motorradwerk, Zschopau starts in 1919 with the commencement of motorcycle production under the DKW name. After the Second World War DKW started again in West Germany, while the Zschopau factory became part of the East German nationalised vehicle manufacturing group, IFA, and its products were sold under this name until the late 1950's when the MZ title came into use. MZ then proceeded to produce a range of road-going motorcycles which were economically priced.

The ETZ range was introduced to supersede the TS range of machines on which they are based. The first ETZ model to be produced was the ETZ250, which appeared in October 1981. It was available in a standard or Luxus (deluxe) form. The standard model had a front drum brake and pre-mix engine lubrication, whereas the Luxus model employed an hydraulic front disc brake and had an oil injection system. Luxus models were also fitted with a tachometer. In August 1983 the standard model was dropped from the range, leaving only the Luxus in production.

In March 1986 the ETZ125 and 150 were introduced, the 150 model basically being a bored out 125. The 125 was available in both standard and Luxus versions whereas the 150 was only available in the Luxus form. The differences between the standard and Luxus models were the same as for the 250. The major difference between the 125/150 models and the 250 was in the use of a totally different engine/gearbox unit. The fuel tank and sidepanels were also redesigned to update the styling of the machine.

In July 1989 the ETZ251 was introduced to replace the 250 model. The 251 is basically an updated 250 with only minor design changes. The main changes being cosmetic which include the tank and sidepanels being restyled to the same shape as the 125 and 150 models.

The UK importers produced an ETZ300 model from July 1987 to 1990. This model was basically a 250 Luxus with a larger bore and a forged alloy piston. A heat dissipater was fitted to the top of the cylinder head. Cosmetically the only change was in the fitting of a handlebar fairing. During 1988 they also produced a customised version of the 125 and 250 Luxus models, known as the ETZ125 XR and the ETZ250 XR. Both models are mechanically identical to the Luxus, the only changes being the fitting of a handlebar fairing, colour-coded front mudguard and a heat dissipater on top of the cylinder head. From late 1989 the importers also replaced the front drum brake on all ETZ125 standard models with a disc brake; this later became original equipment.

Details of the 1991-on models will be found in Chapter 7.

Ordering spare parts

When new parts are required it is advisable to deal direct with a MZ dealer. He is in the best position to offer specialist advice and will be able to supply the more commonly used parts from stock. If the parts need to be ordered, remember that an authorized dealer will be able to arrange faster delivery than a non-specialist dealer. Try to order parts well in advance where this is possible. For example, read through the appropriate section of the manual and see whether gaskets or seals will be needed. This can often avoid having the machine off the road for a week or two while they are ordered.

When ordering, always quote the machine details in full. This will ensure that the correct parts are supplied and will take into account any retrospective manufacturer's modifications. The frame number is found stamped on the right-hand side of the steering head, just in front of the fuel tank. The engine number is stamped on the right-hand side of the crankcase, beside the carburettor.

It is advisable to purchase all spares from an MZ dealer and use only parts of genuine MZ manufacture, although pattern parts are not widely available for these models. Where alternatives are offered, it is prefer-

able to purchase the genuine MZ component to ensure the machine's reliability and safety. Retain any worn or broken parts until the replacements have been obtained; they are sometimes needed as a pattern to help identify the correct replacement when design changes have taken place during a production run.

During the initial warranty period, and as a general rule, make sure that only genuine MZ parts are used. Fitting non-standard parts may well invalidate the warranty, and more importantly, could prove dangerous. Be particularly wary of pattern safety-related parts such as brake and suspension components. These often resemble the original parts very closely and may even be supplied in counterfeit packaging and sold as genuine items.

Some of the more consumable items, such as spark plugs, bulbs, oils, greases and tyres can be purchased from local sources like accessory shops and motor factors, or from mail order suppliers. Always stick to well-known and reputable brands and make sure that the items supplied are suitable for your machine.

Engine number location

Frame number location

Safety first!

Professional motor mechanics are trained in safe working procedures. However enthusiastic you may be about getting on with the job in hand, do take the time to ensure that your safety is not put at risk. A moment's lack of attention can result in an accident, as can failure to observe certain elementary precautions.

There will always be new ways of having accidents, and the following points do not pretend to be a comprehensive list of all dangers; they are intended rather to make you aware of the risks and to encourage a safety-conscious approach to all work you carry out on your vehicle.

Essential DOs and DON'Ts

DON'T start the engine without first ascertaining that the transmission is in neutral.

DON'T suddenly remove the filler cap from a hot cooling system – cover it with a cloth and release the pressure gradually first, or you may get scalded by escaping coolant.

DON'T attempt to drain oil until you are sure it has cooled sufficiently to avoid scalding you.

DON'T grasp any part of the engine, exhaust or silencer without first ascertaining that it is sufficiently cool to avoid burning you.

DON'T allow brake fluid or antifreeze to contact the machine's paintwork or plastic components.

DON'T syphon toxic liquids such as fuel, brake fluid or antifreeze by mouth, or allow them to remain on your skin.

DON'T inhale dust – it may be injurious to health (see *Asbestos* heading).

DON'T allow any spilt oil or grease to remain on the floor – wipe it up straight away, before someone slips on it.

DON'T use ill-fitting spanners or other tools which may slip and cause injury.

DON'T attempt to lift a heavy component which may be beyond your capability – get assistance.

DON'T rush to finish a job, or take unverified short cuts.

DON'T allow children or animals in or around an unattended vehicle.

DON'T inflate a tyre to a pressure above the recommended maximum. Apart from overstressing the carcase and wheel rim, in extreme cases the tyre may blow off forcibly.

DO ensure that the machine is supported securely at all times. This is especially important when the machine is blocked up to aid wheel or fork removal.

DO take care when attempting to slacken a stubborn nut or bolt. It is generally better to pull on a spanner, rather than push, so that if slippage occurs you fall away from the machine rather than on to it.

DO wear eye protection when using power tools such as drill, sander, bench grinder etc.

DO use a barrier cream on your hands prior to undertaking dirty jobs – it will protect your skin from infection as well as making the dirt easier to remove afterwards; but make sure your hands aren't left slippery. Note that long-term contact with used engine oil can be a health hazard.

DO keep loose clothing (cuffs, tie etc) and long hair well out of the way of moving mechanical parts.

DO remove rings, wristwatch etc, before working on the vehicle – especially the electrical system.

DO keep your work area tidy – it is only too easy to fall over articles left lying around.

DO exercise caution when compressing springs for removal or installation. Ensure that the tension is applied and released in a controlled manner, using suitable tools which preclude the possibility of the spring escaping violently.

DO ensure that any lifting tackle used has a safe working load rating adequate for the job.

DO get someone to check periodically that all is well, when working alone on the vehicle.

DO carry out work in a logical sequence and check that everything is correctly assembled and tightened afterwards.

DO remember that your vehicle's safety affects that of yourself and others. If in doubt on any point, get specialist advice.

IF, in spite of following these precautions, you are unfortunate enough to injure yourself, seek medical attention as soon as possible.

Asbestos

Certain friction, insulating, sealing, and other products – such as brake linings, clutch linings, gaskets, etc – contain asbestos. *Extreme care must be taken to avoid inhalation of dust from such products since it is hazardous to health.* If in doubt, assume that they *do* contain asbestos.

Fire

Remember at all times that petrol (gasoline) is highly flammable. Never smoke, or have any kind of naked flame around, when working on the vehicle. But the risk does not end there – a spark caused by an electrical short-circuit, by two metal surfaces contacting each other, by careless use of tools, or even by static electricity built up in your body under certain conditions, can ignite petrol vapour, which in a confined space is highly explosive.

Always disconnect the battery earth (ground) terminal before working on any part of the fuel or electrical system, and never risk spilling fuel on to a hot engine or exhaust.

It is recommended that a fire extinguisher of a type suitable for fuel and electrical fires is kept handy in the garage or workplace at all times. Never try to extinguish a fuel or electrical fire with water.

Note: *Any reference to a 'torch' appearing in this manual should always be taken to mean a hand-held battery-operated electric lamp or flashlight. It does **not** mean a welding/gas torch or blowlamp.*

Fumes

Certain fumes are highly toxic and can quickly cause unconsciousness and even death if inhaled to any extent. Petrol (gasoline) vapour comes into this category, as do the vapours from certain solvents such as trichloroethylene. Any draining or pouring of such volatile fluids should be done in a well ventilated area.

When using cleaning fluids and solvents, read the instructions carefully. Never use materials from unmarked containers – they may give off poisonous vapours.

Never run the engine of a motor vehicle in an enclosed space such as a garage. Exhaust fumes contain carbon monoxide which is extremely poisonous; if you need to run the engine, always do so in the open air or at least have the rear of the vehicle outside the workplace.

The battery

Never cause a spark, or allow a naked light, near the vehicle's battery. It will normally be giving off a certain amount of hydrogen gas, which is highly explosive.

Always disconnect the battery earth (ground) terminal before working on the fuel or electrical systems.

If possible, loosen the filler plugs or cover when charging the battery from an external source. Do not charge at an excessive rate or the battery may burst.

Take care when topping up and when carrying the battery. The acid electrolyte, even when diluted, is very corrosive and should not be allowed to contact the eyes or skin.

If you ever need to prepare electrolyte yourself, always add the acid slowly to the water, and never the other way round. Protect against splashes by wearing rubber gloves and goggles.

Mains electricity and electrical equipment

When using an electric power tool, inspection light etc, always ensure that the appliance is correctly connected to its plug and that, where necessary, it is properly earthed (grounded). Do not use such appliances in damp conditions and, again, beware of creating a spark or applying excessive heat in the vicinity of fuel or fuel vapour. Also ensure that the appliances meet the relevant national safety standards.

Ignition HT voltage

A severe electric shock can result from touching certain parts of the ignition system, such as the HT leads, when the engine is running or being cranked, particularly if components are damp or the insulation is defective. Where an electronic ignition system is fitted, the HT voltage is much higher and could prove fatal.

Tools and working facilities

The first priority when undertaking maintenance or repair work of any sort on a motorcycle is to have a clean, dry, well-lit working area. Work carried out in peace and quiet in the well-ordered atmosphere of a good workshop will give more satisfaction and much better results than can usually be achieved in poor working conditions. A good workshop must have a clean flat workbench or a solidly constructed table of convenient working height. The workbench or table should be equipped with a vice which has a jaw opening of at least 4 in (100 mm). A set of jaw covers should be made from soft metal such as aluminium alloy or copper, or from wood. These covers will minimise the marking or damaging of soft or delicate components which may be clamped in the vice. Some clean, dry, storage space will be required for tools, lubricants and dismantled components. It will be necessary during a major overhaul to lay out engine/gearbox components for examination and to keep them where they will remain undisturbed for as long as is necessary. To this end it is recommended that a supply of metal or plastic containers of suitable size is collected. A supply of clean, lint-free, rags for cleaning purposes and some newspapers, other rags, or paper towels for mopping up spillages should also be kept. If working on a hard concrete floor note that both the floor and one's knees can be protected from oil spillages and wear by cutting open a large cardboard box and spreading it flat on the floor under the machine or workbench. This also helps to provide some warmth in winter and to prevent the loss of nuts, washers, and other tiny components which have a tendency to disappear when dropped on anything other than a perfectly clean, flat, surface.

Unfortunately, such working conditions are not always available to the home mechanic. When working in poor conditions it is essential to take extra time and care to ensure that the components being worked on are kept scrupulously clean and to ensure that no components or tools are lost or damaged.

A selection of good tools is a fundamental requirement for anyone contemplating the maintenance and repair of a motor vehicle. For the owner who does not possess any, their purchase will prove a considerable expense, offsetting some of the savings made by doing-it-yourself. However, provided that the tools purchased meet the relevant national safety standards and are of good quality, they will last for many years and prove an extremely worthwhile investment.

To help the average owner to decide which tools are needed to carry out the various tasks detailed in this manual, we have compiled three lists of tools under the following headings: *Maintenance and minor repair, Repair and overhaul,* and *Specialized.* The newcomer to practical mechanics should start off with the simpler jobs around the vehicle. Then, as his confidence and experience grow, he can undertake more difficult tasks, buying extra tools as and when they are needed. In this way, a *Maintenance and minor repair* tool kit can be built-up into a *Repair and overhaul* tool kit over a considerable period of time without any major cash outlays. The experienced home mechanic will have a tool kit good enough for most repair and overhaul procedures and will add tools from the specialized category when he feels the expense is justified by the amount of use these tools will be put to.

It is obviously not possible to cover the subject of tools fully here. For those who wish to learn more about tools and their use there is a book entitled *Motorcycle Workshop Practice Manual* (Book no 1454) available from the publishers of this manual.

As a general rule, it is better to buy the more expensive, good quality tools. Given reasonable use, such tools will last for a very long time, whereas the cheaper, poor quality, items will wear out faster and need to be renewed more often, thus nullifying the original saving. There is also the risk of a poor quality tool breaking while in use, causing personal injury or expensive damage to the component being worked on.

For practically all tools, a tool factor is the best source since he will have a very comprehensive range compared with the average garage or accessory shop. Having said that, accessory shops often offer excellent quality tools at discount prices, so it pays to shop around. There are plenty of good tools around at reasonable prices, but always aim to purchase items which meet the relevant national safety standards. If in doubt, seek the advice of the shop proprietor or manager before making a purchase.

The basis of any toolkit is a set of spanners. While open-ended spanners with their slim jaws, are useful for working on awkwardly-positioned nuts, ring spanners have advantages in that they grip the nut far more positively. There is less risk of the spanner slipping off the nut and damaging it, for this reason alone ring spanners are preferred. Ideally, the home mechanic should acquire a set of each, but if expense rules this out a set of combination spanners (open-ended at one end and with a ring of the same size at the other) will provide a good compromise. Another item which is so useful it should be considered an essential requirement for any home mechanic is a set of socket spanners. These are available in a variety of drive sizes. It is recommended that the ½-inch drive type is purchased to begin with as although bulkier and more expensive than the ⅜-inch type, the larger size is far more common and will accept a greater variety of torque wrenches, extension pieces and socket sizes. The socket set should comprise sockets of sizes between 8 and 24 mm, a reversible ratchet drive, an extension bar of about 10 inches in length, a spark plug socket with a rubber insert, and a universal joint. Other attachments can be added to the set at a later date.

Maintenance and minor repair tool kit

Set of spanners 8 – 24 mm
Set of sockets and attachments
Spark plug spanner with rubber insert – 14 mm
Adjustable spanner
C-spanner/pin spanner
Torque wrench (same size drive as sockets)
Set of screwdrivers (flat blade)
Set of screwdrivers (cross-head)
Set of Allen keys 4 – 10 mm
Impact screwdriver and bits
Ball pein hammer – 2 lb
Hacksaw (junior)
Self-locking pliers – Mole grips or vice grips
Pliers – combination
Pliers – needle nose
Wire brush (small)
Soft-bristled brush
Tyre pump
Tyre pressure gauge
Tyre tread depth gauge
Oil can
Fine emery cloth
Funnel (medium size)
Drip tray
Grease gun
Set of feeler gauges
Brake bleeding kit
Continuity tester (dry battery and bulb)
Soldering iron and solder
Wire stripper or craft knife
PVC insulating tape
Assortment of split pins, nuts, bolts, and washers

Repair and overhaul toolkit

The tools in this list are virtually essential for anyone undertaking major repairs to a motorcycle and are additional to the tools listed above.

> Plastic or rubber soft-faced mallet
> Pliers – electrician's side cutters
> Circlip pliers – internal (straight or right-angled tips are available)
> Circlip pliers – external
> Cold chisel
> Centre punch
> Pin punch
> Scriber
> Scraper (made from soft metal such as aluminium or copper)
> Soft metal drift
> Steel rule/straightedge
> Assortment of files
> Electric drill and bits
> Wire brush (large)
> Soft wire brush (similar to those used for cleaning suede shoes)
> Sheet of plate glass
> Hacksaw (large)
> Stud extractor set (E-Z out)

Specialized tools

This is not a list of the tools made by the machine's manufacturer to carry out a specific task on a limited range of models. Occasional references are made to such tools in the text of this manual and, in general, an alternative method of carrying out the task without the manufacturer's tool is given where possible. The tools mentioned in this list are those which are not used regularly and are expensive to buy in view of their infrequent use. Where this is the case it may be possible to hire or borrow the tools against a deposit from a local dealer or tool hire shop. An alternative is for a group of friends or a motorcycle club to join in the purchase.

> Piston ring compressor
> Universal bearing puller
> Cylinder bore honing attachment (for electric drill)
> Micrometer set
> Vernier calipers
> Dial gauge set
> Multimeter
> Dwell meter/tachometer

Care and maintenance of tools

Whatever the quality of the tools purchased, they will last much longer if cared for. This means in practice ensuring that a tool is used for its intended purpose; for example screwdrivers should not be used as a substitute for a centre punch, or as chisels. Always remove dirt or grease and any metal particles but remember that a light film of oil will prevent rusting if the tools are infrequently used. The common tools can be kept together in a large box or tray but the more delicate, and more expensive, items should be stored separately where they cannot be damaged. When a tool is damaged or worn out, be sure to renew it immediately. It is false economy to continue to use a worn spanner or screwdriver which may slip and cause expensive damage to the component being worked on.

Fastening systems

Fasteners, basically, are nuts, bolts and screws used to hold two or more parts together. There are a few things to keep in mind when working with fasteners. Almost all of them use a locking device of some type; either a lock washer, locknut, locking tab or thread adhesive. All threaded fasteners should be clean, straight, have undamaged threads and undamaged corners on the hexagon head where the spanner fits. Develop the habit of replacing all damaged nuts and bolts with new ones.

Rusted nuts and bolts should be treated with a rust penetrating fluid to ease removal and prevent breakage. After applying the rust penetrant, let it 'work' for a few minutes before trying to loosen the nut or bolt. Badly rusted fasteners may have to be chiselled off or removed with a special nut breaker, available at tool shops.

Flat washers and lock washers, when removed from an assembly should always be replaced exactly as removed. Replace any damaged washers with new ones. Always use a flat washer between a lock washer and any soft metal surface (such as aluminium), thin sheet metal or plastic. Special locknuts can only be used once or twice before they lose their locking ability and must be renewed.

If a bolt or stud breaks off in an assembly, it can be drilled out and removed with a special tool called an E-Z out. Most dealer service departments and motorcycle repair shops can perform this task, as well as others (such as the repair of threaded holes that have been stripped out).

Spanner size comparison

Jaw gap (in)	Spanner size	Jaw gap (in)	Spanner size
0.250	$\frac{1}{4}$ in AF	0.945	24 mm
0.276	7 mm	1.000	1 in AF
0.313	$\frac{5}{16}$ in AF	1.010	$\frac{9}{16}$ in Whitworth; $\frac{5}{8}$ in BSF
0.315	8 mm	1.024	26 mm
0.344	$\frac{11}{32}$ in AF; $\frac{3}{16}$ in Whitworth	1.063	$1\frac{1}{16}$ in AF; 27 mm
0.354	9 mm	1.100	$\frac{5}{8}$ in Whitworth; $\frac{11}{16}$ in BSF
0.375	$\frac{3}{8}$ in AF	1.125	$1\frac{1}{8}$ in AF
0.394	10 mm	1.181	30 mm
0.433	11 mm	1.200	$\frac{11}{16}$ in Whitworth; $\frac{3}{4}$ in BSF
0.438	$\frac{7}{16}$ in AF	1.250	$1\frac{1}{4}$ in AF
0.445	$\frac{7}{32}$ in Whitworth; $\frac{1}{4}$ in BSF	1.260	32 mm
0.472	12 mm	1.300	$\frac{3}{4}$ in Whitworth; $\frac{7}{8}$ in BSF
0.500	$\frac{1}{2}$ in AF	1.313	$1\frac{5}{16}$ in AF
0.512	13 mm	1.390	$\frac{13}{16}$ in Whitworth; $\frac{15}{16}$ in BSF
0.525	$\frac{1}{4}$ in Whitworth; $\frac{5}{16}$ in BSF	1.417	36 mm
0.551	14 mm	1.438	$1\frac{7}{16}$ in AF
0.563	$\frac{9}{16}$ in AF	1.480	$\frac{7}{8}$ in Whitworth; 1 in BSF
0.591	15 mm	1.500	$1\frac{1}{2}$ in AF
0.600	$\frac{5}{16}$ in Whitworth; $\frac{3}{8}$ in BSF	1.575	40 mm; $1\frac{5}{16}$ in Whitworth
0.625	$\frac{5}{8}$ in AF	1.614	41 mm
0.630	16 mm	1.625	$1\frac{5}{8}$ in AF
0.669	17 mm	1.670	1 in Whitworth; $1\frac{1}{8}$ in BSF
0.686	$1\frac{1}{16}$ in AF	1.688	$1\frac{11}{16}$ in AF
0.709	18 mm	1.811	46 mm
0.710	$\frac{3}{8}$ in Whitworth; $\frac{7}{16}$ in BSF	1.813	$1\frac{13}{16}$ in AF
0.748	19 mm	1.860	$1\frac{1}{8}$ in Whitworth; $1\frac{1}{4}$ in BSF
0.750	$\frac{3}{4}$ in AF	1.875	$1\frac{7}{8}$ in AF
0.813	$1\frac{3}{16}$ in AF	1.969	50 mm
0.820	$\frac{7}{16}$ in Whitworth; $\frac{1}{2}$ in BSF	2.000	2 in AF
0.866	22 mm	2.050	$1\frac{1}{4}$ in Whitworth; $1\frac{3}{8}$ in BSF
0.875	$\frac{7}{8}$ in AF	2.165	55 mm
0.920	$\frac{1}{2}$ in Whitworth; $\frac{9}{16}$ in BSF	2.362	60 mm
0.938	$1\frac{5}{16}$ in AF		

Standard torque settings

Specific torque settings will be found at the end of the specifications section of each chapter. Where no figure is given, it should be secured according to the table below.

Fastener type (thread diameter)	kgf m	lbf ft
5 mm bolt or nut	0.45 – 0.6	3.5 – 4.5
6 mm bolt or nut	0.8 – 1.2	6 – 9
8 mm bolt or nut	1.8 – 2.5	13 – 18
10 mm bolt or nut	3.0 – 4.0	22 – 29
12 mm bolt or nut	5.0 – 6.0	36 – 43
5 mm screw	0.35 – 0.5	2.5 – 3.6
6 mm screw	0.7 – 1.1	5 – 8
6 mm flange bolt	1.0 – 1.4	7 – 10
8 mm flange bolt	2.4 – 3.0	17 – 22
10 mm flange bolt	3.5 – 4.5	25 – 33

Choosing and fitting accessories

The range of accessories available to the modern motorcyclist is almost as varied and bewildering as the range of motorcycles. This Section is intended to help the owner in choosing the correct equipment for his needs and to avoid some of the mistakes made by many riders when adding accessories to their machines. It will be evident that the Section can only cover the subject in the most general terms and so it is recommended that the owner, having decided that he wants to fit, for example, a luggage rack or carrier, seeks the advice of several local dealers and the owners of similar machines. This will give a good idea of what makes of carrier are easily available, and at what price. Talking to other owners will give some insight into the drawbacks or good points of any one make. A walk round the motorcycles in car parks or outside a dealer will often reveal the same sort of information.

The first priority when choosing accessories is to assess exactly what one needs. It is, for example, pointless to buy a large heavy-duty carrier which is designed to take the weight of fully laden panniers and topbox when all you need is a place to strap on a set of waterproofs and a lunchbox when going to work. Many accessory manufacturers have ranges of equipment to cater for the individual needs of different riders and this point should be borne in mind when looking through a dealer's catalogues. Having decided exactly what is required and the use to which the accessories are going to be put, the owner will need a few hints on what to look for when making the final choice. To this end the Section is now sub-divided to cover the more popular accessories fitted. Note that it is in no way a customizing guide, but merely seeks to outline the practical considerations to be taken into account when adding aftermarket equipment to a motorcycle.

Fairings and windscreens

A fairing is possibly the single, most expensive, aftermarket item to be fitted to any motorcycle and, therefore, requires the most thought before purchase. Fairings can be divided into two main groups: front fork mounted handlebar fairings and windscreens, and frame mounted fairings.

The first group, the front fork mounted fairings, are becoming far more popular than was once the case, as they offer several advantages over the second group. Front fork mounted fairings generally are much easier and quicker to fit, involve less modification to the motorcycle, do not as a rule restrict the steering lock, permit a wider selection of handlebar styles to be used, and offer adequate protection for much less money than the frame mounted type. They are also lighter, can be swapped easily between different motorcycles, and are available in a much greater variety of styles. Their main disadvantages are that they do not offer as much weather protection as the frame mounted types, rarely offer any storage space, and, if poorly fitted or naturally incompatible, can have an adverse effect on the stability of the motorcycle.

The second group, the frame mounted fairings, are secured so rigidly to the main frame of the motorcycle that they can offer a substantial amount of protection to motorcycle and rider in the event of a crash. They offer almost complete protection from the weather and, if double-skinned in construction, can provide a great deal of useful storage space. The feeling of peace, quiet and complete relaxation encountered when riding behind a good full fairing has to be experienced to be believed. For this reason full fairings are considered essential by most touring motorcyclists and by many people who ride all year round. The main disadvantages of this type are that fitting can take a long time, often involving removal or modification of standard motorcycle components, they restrict the steering lock and they can add up to about 40 lb to the weight of the machine. They do not usually affect the stability of the machine to any great extent once the front tyre

pressure and suspension have been adjusted to compensate for the extra weight, but can be affected by sidewinds.

The first thing to look for when purchasing a fairing is the quality of the fittings. A good fairing will have strong, substantial brackets constructed from heavy-gauge tubing; the brackets must be shaped to fit the frame or forks evenly so that the minimum of stress is imposed on the assembly when it is bolted down. The brackets should be properly painted or finished – a nylon coating being the favourite of the better manufacturers – the nuts and bolts provided should be of the same thread and size standard as is used on the motorcycle and be properly plated. Look also for shakeproof locking nuts or locking washers to ensure that everything remains securely tightened down. The fairing shell is generally made from one of two materials: fibreglass or ABS plastic. Both have their advantages and disadvantages, but the main consideration for the owner is that fibreglass is much easier to repair in the event of damage occurring to the fairing. Whichever material is used, check that it is properly finished inside as well as out, that the edges are protected by beading and that the fairing shell is insulated from vibration by the use of rubber grommets at all mounting points. Also be careful to check that the windscreen is retained by plastic bolts which will snap on impact so that the windscreen will break away and not cause personal injury in the event of an accident.

Having purchased your fairing or windscreen, read the manufacturer's fitting instructions very carefully and check that you have all the necessary brackets and fittings. Ensure that the mounting brackets are located correctly and bolted down securely. Note that some manufacturers use hose clamps to retain the mounting brackets; these should be discarded as they are convenient to use but not strong enough for the task. Stronger clamps should be substituted; car exhaust pipe clamps of suitable size would be a good alternative. Ensure that the front forks can turn through the full steering lock available without fouling the fairing. With many types of frame-mounted fairing the handlebars will have to be altered or a different type fitted and the steering lock will be restricted by stops provided with the fittings. Also check that the fairing does not foul the front wheel or mudguard, in any steering position, under full fork compression. Re-route any cables, brake pipes or electrical wiring which may snag on the fairing and take great care to protect all electrical connections, using insulating tape. If the manufacturer's instructions are followed carefully at every stage no serious problems should be encountered. Remember that hydraulic pipes that have been disconnected must be carefully re-tightened and the hydraulic system purged of air bubbles by bleeding.

Two things will become immediately apparent when taking a motorcycle on the road for the first time with a fairing – the first is the tendency to underestimate the road speed because of the lack of wind pressure on the body. This must be very carefully watched until one has grown accustomed to riding behind the fairing. The second thing is the alarming increase in engine noise which is an unfortunate but inevitable by-product of fitting any type of fairing or windscreen, and is caused by normal engine noise being reflected, and in some cases amplified, by the flat surface of the fairing.

Luggage racks or carriers

Carriers are possibly the commonest item to be fitted to modern motorcycles. They vary enormously in size, carrying capacity, and durability. When selecting a carrier, always look for one which is made specifically for your machine and which is bolted on with as few separate brackets as possible. The universal-type carrier, with its mass of brackets and adaptor pieces, will generally prove too weak to be of any real use. A good carrier should bolt to the main frame, generally using the two suspension unit top mountings and a mudguard mounting bolt as attachment points, and have its luggage platform as low and as far forward as possible to minimise the effect of any load on the machine's stability. Look for good quality, heavy gauge tubing, good welding and good finish. Also ensure that the carrier does not prevent opening of the seat, sidepanels or tail compartment, as appropriate. When using a carrier, be very careful not to overload it. Excessive weight placed so high and so far to the rear of any motorcycle will have an adverse effect on the machine's steering and stability.

Luggage

Motorcycle luggage can be grouped under two headings: soft and hard. Both types are available in many sizes and styles and have advantages and disadvantages in use.

Soft luggage is now becoming very popular because of its lower cost and its versatility. Whether in the form of tankbags, panniers, or strap-on bags, soft luggage requires in general no brackets and no modification to the motorcycle. Equipment can be swapped easily from one motorcycle to another and can be fitted and removed in seconds. Awkwardly shaped loads can easily be carried. The disadvantages of soft luggage are that the contents cannot be secure against the casual thief, very little protection is afforded in the event of a crash, and waterproofing is generally poor. Also, in the case of panniers, carrying capacity is restricted to approximately 10 lb, although this amount will vary considerably depending on the manufacturer's recommendation. When purchasing soft luggage, look for good quality material, generally vinyl or nylon, with strong, well-stitched attachment points. It is always useful to have separate pockets, especially on tank bags, for items which will be needed on the journey. When purchasing a tank bag, look for one which has a separate, well-padded, base. This will protect the tank's paintwork and permit easy access to the filler cap at petrol stations.

Hard luggage is confined to two types: panniers, and top boxes or tail trunks. Most hard luggage manufacturers produce matching sets of these items, the basis of which is generally that manufacturer's own heavy-duty luggage rack. Variations on this theme occur in the form of separate frames for the better quality panniers, fixed or quickly-detachable luggage, and in size and carrying capacity. Hard luggage offers a reasonable degree of security against theft and good protection against weather and accident damage. Carrying capacity is greater than that of soft luggage, around 15 – 20 lb in the case of panniers, although top boxes should never be loaded as much as their apparent capacity might imply. A top box should only be used for lightweight items, because one that is heavily laden can have a serious effect on the stability of the machine. When purchasing hard luggage look for the same good points as mentioned under fairings and windscreens, ie good quality mounting brackets and fittings, and well-finished fibreglass or ABS plastic cases. Again as with fairings, always purchase luggage made specifically for your motorcycle, using as few separate brackets as possible, to ensure that everything remains securely bolted in place. When fitting hard luggage, be careful to check that the rear suspension and brake operation will not be impaired in any way and remember that many pannier kits require re-siting of the indicators. Remember also that a non-standard exhaust system may make fitting extremely difficult.

Handlebars

The occupation of fitting alternative types of handlebar is extremely popular with modern motorcyclists, whose motives may vary from the purely practical, wishing to improve the comfort of their machines, to the purely aesthetic, where form is more important than function. Whatever the reason, there are several considerations to be borne in mind when changing the handlebars of your machine. If fitting lower bars, check carefully that the switches and cables do not foul the petrol tank on full lock and that the surplus length of cable, brake pipe, and electrical wiring are smoothly and tidily disposed of. Avoid tight kinks in cable or brake pipes which will produce stiff controls or the premature and disastrous failure of an overstressed component. If necessary, remove the petrol tank and re-route the cable from the engine/gearbox unit upwards, ensuring smooth gentle curves are produced. In extreme cases, it will be necessary to purchase a shorter brake pipe to overcome this problem. In the case of higher handlebars than standard it will almost certainly be necessary to purchase extended cables and brake pipes. Fortunately, many standard motorcycles have a custom version which will be equipped with higher handlebars and, therefore, factory-built extended components will be available from your local dealer. It is not usually necessary to extend electrical wiring, as switch clusters may be used on several different motorcycles, some being custom versions. This point should be borne in mind however when fitting extremely high or wide handlebars.

When fitting different types of handlebar, ensure that the mounting clamps are correctly tightened to the manufacturer's specifications and that cables and wiring, as previously mentioned, have smooth easy runs and do not snag on any part of the motorcycle throughout the full steering lock. Ensure that the fluid level in the front brake master cylinder remains level to avoid any chance of air entering the hydraulic system. Also check that the cables are adjusted correctly and that all handlebar controls operate correctly and can be easily reached when riding.

Crashbars

Crashbars, also known as engine protector bars, engine guards, or case savers, are extremely useful items of equipment which can contribute protection to the machine's structure if a crash occurs. They do not, as has been inferred in the US, prevent the rider from crashing, or necessarily prevent rider injury should a crash occur.

It is recommended that only the smaller, neater, engine protector type of crashbar is considered. This type will offer protection while restricting, as little as is possible, access to the engine and the machine's ground clearance. The crashbars should be designed for use specifically on your machine, and should be constructed of heavy-gauge tubing with strong, integral mounting brackets. Where possible, they should bolt to a strong lug on the frame, usually at the engine mounting bolts.

The alternative type of crashbar is the larger cage type. This type is not recommended in spite of their appearance which promises some protection to the rider as well as to the machine. The larger amount of leverage imposed by the size of this type of crashbar increases the risk of severe frame damage in the event of an accident. This type also decreases the machine's ground clearance and restricts access to the engine. The amount of protection afforded the rider is open to some doubt as the design is based on the premise that the rider will stay in the normally seated position during an accident, and the crash bar structure will not itself fail. Neither result can in any way be guaranteed.

As a general rule, always purchase the best, ie usually the most expensive, set of crashbars you an afford. The investment will be repaid by minimising the amount of damage incurred, should the machine be involved in an accident. Finally, avoid the universal type of crashbar. This should be regarded only as a last resort to be used if no alternative exists. With its usual multitude of separate brackets and spacers, the universal crashbar is far too weak in design and construction to be of any practical value.

Electrical equipment

The vast range of electrical equipment available to motorcyclists is so large and so diverse that only the most general outline can be given here. Electrical accessories vary from electronic ignition kits fitted to replace contact breaker points, to additional lighting at the front and rear, more powerful horns, various instruments and gauges, clocks, anti-theft systems, heated clothing, CB radios, radio-cassette players, and intercom systems, to name but a few of the more popular items of equipment.

As will be evident, it would require a separate manual to cover this subject alone and this section is therefore restricted to outlining a few basic rules which must be borne in mind when fitting electrical equipment. The first consideration is whether your machine's electrical system has enough reserve capacity to cope with the added demand of the accessories you wish to fit. The motorcycle's manufacturer or importer should be able to furnish this sort of information and may also be able to offer advice on uprating the electrical system. Failing this, a good dealer or the accessory manufacturer may be able to help. In some cases, more powerful generator components may be available, perhaps from another motorcycle in the manufacturer's range. The second consideration is the legal requirements in force in your area. The local police may be prepared to help with this point. In the UK for example, there are strict regulations governing the position and use of auxiliary riding lamps and fog lamps.

When fitting electrical equipment always disconnect the battery first to prevent the risk of a short-circuit, and be careful to ensure that all connections are properly made and that they are waterproof. Remember that many electrical accessories are designed primarily for use in cars and that they cannot easily withstand the exposure to vibration and to the weather. Delicate components must be rubber-mounted to insulate them from vibration, and sealed carefully to prevent the entry of rainwater and dirt. Be careful to follow exactly the accessory manufacturer's instructions in conjunction with the wiring diagram at the back of this manual.

Accessories – general

Accessories fitted to your motorcycle will rapidly deteriorate if not cared for. Regular washing and polishing will maintain the finish and will provide an opportunity to check that all mounting bolts and nuts are securely fastened. Any signs of chafing or wear should be watched for, and the cause cured as soon as possible before serious damage occurs.

As a general rule, do not expect the re-sale value of your motorcycle to increase by an amount proportional to the amount of money and effort put into fitting accessories. It is usually the case that an absolutely standard motorcycle will sell more easily at a better price than one that has been modified. If you are in the habit of exchanging your machine for another at frequent intervals, this factor should be borne in mind to avoid loss of money.

Fault diagnosis

Contents

1 Introduction

This Section provides an easy reference-guide to the more common faults that are likely to afflict your machine. Obviously, the opportunities are almost limitless for faults to occur as a result of obscure failures, and to try and cover all eventualities would require a book. Indeed, a number have been written on the subject.

Successful fault diagnosis is not a mysterious 'black art' but the application of a bit of knowledge combined with a systematic and logical approach to the problem. Approach any fault diagnosis by first accurately identifying the symptom and then checking through the list of possible causes, starting with the simplest or most obvious and progressing in stages to the most complex. Take nothing for granted, but above all apply liberal quantities of common sense.

The main symptom of a fault is given in the text as a major heading below which are listed, as Section headings, the various systems or areas which may contain the fault. Details of each possible cause for a fault and the remedial action to be taken are given, in brief, in the paragraphs below each Section heading. Further information should be sought in the relevant Chapter.

Engine does not start when turned over

2 No fuel flow to carburettor

● Fuel tank empty or level too low. Check that the tap is turned to 'On' or 'Reserve' position as required. If in doubt, prise off the fuel feed pipe at the carburettor end and check that fuel runs from the pipe when the tap is turned on.

● Tank filler cap vent obstructed. This can prevent fuel from flowing into the carburettor float bowl because air cannot enter the fuel tank to replace it. The problem is more likely to appear when the machine is being ridden. Check by listening close to the filler cap and releasing it. A hissing noise indicates that a blockage is present. Remove the cap and clear the vent hole with wire or by using an air line from the inside of the cap.

● Fuel tap or filter blocked. Blockage may be due to accumulation of rust or paint flakes from the tank's inner surface or of foreign matter from contaminated fuel. Remove the tap and clean it and the filter. Look also for water droplets in the fuel.

● Fuel line blocked. Blockage of the fuel line is more likely to result from a kink in the line rather than the accumulation of debris.

3 Fuel not reaching cylinder

● Float chamber not filling. Caused by float needle or floats sticking in up position. This may occur after the machine has been left standing for an extended length of time allowing the fuel to evaporate. When this occurs a gummy residue is often left which hardens to a varnish-like substance. This condition may be worsened by corrosion and crystalline deposits produced prior to the total evaporation of contaminated fuel. Sticking of the float needle may also be caused by wear. In any case removal of the float chamber will be necessary for inspection and cleaning.

● Blockage in starting circuit, slow running circuit or jets. Blockage of these items may be attributable to debris from the fuel tank by-passing the filter system or to gumming up as described in paragraph 1. Water droplets in the fuel will also block jets and passages. The carburettor should be dismantled for cleaning.

● Fuel level too low. The fuel level in the float chamber is controlled by float height. The fuel level may increase with wear or damage but will never reduce, thus a low fuel level is an inherent rather than developing condition. Check the float height, renewing the float or needle if required.

● Oil blockage in fuel system or carburettor (petroil lubricated engines only). May arise when the machine has been parked for long periods and the residual petrol has evaporated. To rectify, dismantle and clean the carburettor and tap, flush the tank and fill with fresh petroil mixed in the correct proportions. This problem can be avoided by running the float bowl dry before the machine is stored for long periods. Do not attempt to use fuel which has become stale.

4 Engine flooding

● Float valve needle worn or stuck open. A piece of rust or other debris can prevent correct seating of the needle against the valve seat thereby permitting an uncontrolled flow of fuel. Similarly, a worn needle or needle seat will prevent valve closure. Dismantle the carburettor float bowl for cleaning and, if necessary, renewal of the worn components.

● Fuel level too high. The fuel level is controlled by the float height which may increase due to wear of the float needle, pivot pin or operating tang. Check the float height, and make any necessary adjustments. A leaking float will cause an increase in fuel level, and thus should be renewed.

● Cold starting mechanism. Check the choke (starter mechanism) for correct operation. If the mechanism jams in the 'On' position subsequent starting of a hot engine will be difficult.

● Blocked air filter. A badly restricted air filter will cause flooding. Check the filter and clean or renew as required. A collapsed inlet hose will have a similar effect. Check that the air filter inlet has not become blocked by a rag or similar item.

5 No spark at plug

● Ignition switch not on.
● Fuse blown. Check fuse for ignition circuit. See wiring diagram.
● Spark plug dirty, oiled or 'whiskered'. Because the induction mixture

of a two-stroke engine is inclined to be of a rather oily nature it is comparatively easy to foul the plug electrodes, especially where there have been repeated attempts to start the engine. A machine used for short journeys will be more prone to fouling because the engine may never reach full operating temperature, and the deposits will not burn off. On rare occasions a change of plug grade may be required but the advice of a dealer should be sought before making such a change. 'Whiskering' is a comparatively rare occurrence on modern machines but may be encountered where pre-mixed petrol and oil (petroil) lubrication is employed. An electrode deposit in the form of a barely visible filament across the plug electrodes can short circuit the plug and prevent its sparking. On all two-stroke machines it is a sound precaution to carry a new spare spark plug for substitution in the event of fouling problems.

● Spark plug failure. Clean the spark plug thoroughly and reset the electrode gap. Refer to the spark plug section and the colour condition guide in Routine maintenance. If the spark plug shorts internally or has sustained visible damage to the electrodes, core or ceramic insulator it should be renewed. On rare occasions a plug that appears to spark vigorously will fail to do so when refitted to the engine and subjected to the compression pressure in the cylinder.

● Spark plug cap or high tension (HT) lead faulty. Check condition and security. Replace if deterioration is evident. Most spark plug caps have an internal resistor designed to inhibit electrical interference with radio and television sets. On rare occasions the resistor may break down, thus preventing sparking. If this is suspected, fit a new cap as a precaution.

● Spark plug cap loose. Check that the spark plug cap fits securely over the plug and, where fitted, the screwed terminal on the plug end is secure.

● Shorting due to moisture. Certain parts of the ignition system are susceptible to shorting when the machine is ridden or parked in wet weather. Check particularly the area from the spark plug cap back to the ignition coil. A water dispersant spray may be used to dry out water-logged components. Recurrence of the problem can be prevented by using an ignition sealant spray after drying out and cleaning.

● Ignition switch shorted. May be caused by water corrosion or wear. Water dispersant and contact cleaning sprays may be used. If this fails to overcome the problem dismantling and visual inspection of the switches will be required.

● Shorting or open circuit in wiring. Failure in any wire connecting any of the ignition components will cause ignition malfunction. Check also that all connections are clean, dry and tight.

● Ignition coil failure. Check the coil, referring to Chapter 3.

● Capacitor (condenser) failure. The capacitor may be checked most easily by substitution with a replacement item. Blackened contact breaker points indicate capacitor malfunction but this may not always occur.

● Contact breaker points pitted, burned or closed up. Check the contact breaker points, referring to Routine maintenance. Check also that the low tension leads at the contact breaker are secure and not shorting out.

6 Weak spark at plug

● Feeble sparking at the plug may be caused by any of the faults mentioned in the preceding Section other than those items in the first three paragraphs. Check first the contact breaker assembly and the spark plug, these being the most likely culprits.

7 Compression low

● Spark plug loose. This will be self-evident on inspection, and may be accompanied by a hissing noise when the engine is turned over. Remove the plug and check that the threads in the cylinder head are not damaged. Check also that the plug sealing washer is in good condition.

● Cylinder head gasket leaking. This condition is often accompanied by a high pitched squeak from around the cylinder head and oil loss, and may be caused by insufficiently tightened cylinder head fasteners, a warped cylinder head or mechanical failure of the gasket material. Re-torqueing the fasteners to the correct specification may seal the leak in some instances but if damage has occurred this course of action will provide, at best, only a temporary cure.

● Low crankcase compression. This can be caused by worn main bearings and seals and will upset the incoming fuel/air mixture. A good seal in these areas is essential on any two-stroke engine.

● Piston rings sticking or broken. Sticking of the piston rings may be caused by seizure due to lack of lubrication or overheating as a result of poor carburation or incorrect fuel type. Gumming of the rings may result from lack of use, or carbon deposits in the ring grooves. Broken rings result from over-revving, over-heating or general wear. In either case a top-end overhaul will be required.

Engine stalls after starting

8   General causes

● Improper cold start mechanism operation. Check that the operating controls function smoothly and, where applicable, are correctly adjusted. A cold engine may not require application of an enriched mixture to start initially but may baulk without choke once firing. Likewise a hot engine may start with an enriched mixture but will stop almost immediately if the choke is inadvertently in operation.

● Ignition malfunction. See Section 9. Weak spark at plug.

● Carburettor incorrectly adjusted. Maladjustment of the mixture strength or idle speed may cause the engine to stop immediately after starting. See Chapter 2.

● Fuel contamination. Check for filter blockage by debris or water which reduces, but does not completely stop, fuel flow, or blockage of the slow speed circuit in the carburettor by the same agents. If water is present it can often be seen as droplets in the bottom of the float bowl. Clean the filter and, where water is in evidence, drain and flush the fuel tank and float bowl.

● Intake air leak. Check for security of the carburettor mounting and hose connections, and for cracks or splits in the hoses. Check also that the carburettor top is secure.

● Air filter blocked or omitted. A blocked filter will cause an over-rich mixture; the omission of a filter will cause an excessively weak mixture. Both conditions will have a detrimental effect on carburation. Clean or renew the filter as necessary.

● Fuel filler cap air vent blocked. Usually caused by dirt or water. Clean the vent orifice.

● Choked exhaust system. Caused by excessive carbon build-up in the system, particularly around the silencer baffles. In many cases these can be detached for cleaning, though mopeds have one-piece systems which require a rather different approach. Refer to Routine maintenance for further information.

● Excessive carbon build-up in the engine. This can result from failure to decarbonise the engine at the specified interval or through excessive oil consumption. On pump-fed engines check pump adjustment. On pre-mix (petroil) systems check that oil is mixed in the recommended ratio.

Poor running at idle and low speed

9   Weak spark at plug or erratic firing

● Battery voltage low. In certain conditions low battery charge, especially when coupled with a badly sulphated battery, may result in misfiring. If the battery is in good general condition it should be recharged; an old battery suffering from sulphated plates should be renewed.

● Spark plug fouled, faulty or incorrectly adjusted. See Section 4 or refer to Routine maintenance.

● Spark plug cap or high tension lead shorting. Check the condition of both these items ensuring that they are in good condition and dry and that the cap is fitted correctly.

● Spark plug type incorrect. Fit plug of correct type and heat range as given in Specifications. In certain conditions a plug of hotter or colder type may be required for normal running.

● Contact breaker points pitted, burned or closed-up. Check the contact breaker assembly, referring to Routine maintenance.

● Ignition timing incorrect. Check the ignition timing as described in Routine maintenance.

● Faulty ignition coil. Partial failure of the coil internal insulation will

diminish the performance of the coil. No repair is possible, a new component must be fitted.

● Faulty capacitor (condenser). A failure of the capacitor will cause blackening of the contact breaker point faces and will allow excessive sparking at the points. A faulty capacitor may best be checked by substitution of a serviceable replacement item.

● Defective alternator. Refer to Chapter 6 for further details on test procedures.

10   Fuel/air mixture incorrect

● Intake air leak. Check carburettor mountings and air cleaner hoses for security and signs of splitting. Ensure that carburettor top is secure.

● Mixture strength incorrect. Adjust slow running mixture strength using pilot adjustment screw.

● Pilot jet or slow running circuit blocked. The carburettor should be removed and dismantled for thorough cleaning. Blow through all jets and air passages with compressed air to clear obstructions.

● Air cleaner clogged or omitted. Clean or fit air cleaner element as necessary. Check also that the element and air filter cover are correctly seated.

● Cold start mechanism in operation. Check that the choke has not been left on inadvertently and the operation is correct. Where applicable check the operating cable free play.

● Fuel level too high or too low. Check the float height, renewing float or needle if required. See Section 3 or 4.

● Fuel tank air vent obstructed. Obstructions usually caused by dirt or water. Clean vent orifice.

11   Compression low

● See Section 7.

Acceleration poor

12   General causes

● All items as for previous Section.

● Choked air filter. Failure to keep the air filter element clean will allow the build-up of dirt with proportional loss of performance. In extreme cases of neglect acceleration will suffer.

● Choked exhaust system. This can result from failure to remove accumulations of carbon from the silencer baffles at the prescribed intervals. The increased back pressure will make the machine noticeably sluggish. Refer to Routine maintenance for further information on decarbonisation.

● Excessive carbon build-up in the engine. This can result from failure to decarbonise the engine at the specified interval or through excessive oil consumption. On pump-fed engines check pump adjustment. On pre-mix (petroil) systems check that oil is mixed in the recommended ratio.

● Ignition timing incorrect. Check the contact breaker gap and set within the prescribed range ensuring that the ignition timing is correct. If the contact breaker assembly is worn it may prove impossible to get the gap and timing settings to coincide, necessitating renewal.

● Carburation fault. See Section 10.

● Mechanical resistance. Check that the brakes are not binding. On small machines in particular note that the increased rolling resistance caused by under-inflated tyres may impede acceleration.

Poor running or lack of power at high speeds

13   Weak spark at plug or erratic firing

● All items as for Section 9.

● HT lead insulation failure. Insulation failure of the HT lead and spark plug cap due to old age or damage can cause shorting when the engine is driven hard. This condition may be less noticeable, or not noticeable at all at lower engine speeds.

14 Fuel/air mixture incorrect

● All items as for Section 10, with the exception of items relative exclusively to low speed running.
● Main jet blocked. Debris from contaminated fuel, or from the fuel tank, and water in the fuel can block the main jet. Clean the fuel filter, the float bowl area, and if water is present, flush and refill the fuel tank.
● Main jet is the wrong size. The standard carburettor jetting is for sea level atmospheric pressure. For high altitudes, usually above 5000 ft, a smaller main jet will be required.
● Jet needle and needle jet worn. These can be renewed individually but should be renewed as a pair. Renewal of both items requires partial dismantling of the carburettor.
● Air bleed holes blocked. Dismantle carburettor and use compressed air to blow out all air passages.
● Reduced fuel flow. A reduction in the maximum fuel flow from the fuel tank to the carburettor will cause fuel starvation, proportionate to the engine speed. Check for blockages through debris or a kinked fuel line.

15 Compression low

● See Section 7.

Knocking or pinking

16 General causes

● Carbon build-up in combustion chamber. After a high mileage has been covered large accumulations of carbon may occur. This may glow red hot and cause premature ignition of the fuel/air mixture, in advance of normal firing by the spark plug. Cylinder head removal will be required to allow inspection and cleaning.
● Fuel incorrect. A low grade fuel, or one of poor quality may result in compression induced detonation of the fuel resulting in knocking and pinking noises. Old fuel can cause similar problems. A too highly leaded fuel will reduce detonation but will accelerate deposit formation in the combustion chamber and may lead to early pre-ignition as described in item 1.
● Spark plug heat range incorrect. Uncontrolled pre-ignition can result from the use of a spark plug the heat range of which is too hot.
● Weak mixture. Overheating of the engine due to a weak mixture can result in pre-ignition occurring where it would not occur when engine temperature was within normal limits. Maladjustment, blocked jets or passages and air leaks can cause this condition.

Overheating

17 Firing incorrect

● Spark plug fouled, defective or maladjusted. See Section 5.
● Spark plug type incorrect. Refer to the Specifications and ensure that the correct plug type is fitted.
● Incorrect ignition timing. Timing that is far too much advanced or far too much retarded will cause overheating. Check the ignition timing is correct.

18 Fuel/air mixture incorrect

● Slow speed mixture strength incorrect. Adjust pilot air screw.
● Main jet wrong size. The carburettor is jetted for sea level atmospheric conditions. For high altitudes, usually above 5000 ft, a smaller main jet will be required.
● Air filter badly fitted or omitted. Check that the filter element is in

place and that it and the air filter box cover are sealing correctly. Any leaks will cause a weak mixture.
● Induction air leaks. Check the security of the carburettor mountings and hose connections, and for cracks and splits in the hoses. Check also that the carburettor top is secure.
● Fuel level too low. See Section 3.
● Fuel tank filler cap air vent obstructed. Clear blockage.

19 Lubrication inadequate

● Petrol/oil mixture incorrect. The proportion of oil mixed with the petrol in the tank is critical if the engine is to perform correctly. Too little oil will leave the reciprocating parts and bearings poorly lubricated and overheating will occur. In extreme case the engine will seize. Conversely, too much oil will effectively displace a similar amount of petrol. Though this does not often cause overheating in practice it is possible that the resultant weak mixture may cause overheating. It will inevitably cause a loss of power and excessive exhaust smoke.
● Oil pump settings incorrect. The oil pump settings are of great importance since the quantities of oil being injected are very small. Any variation in oil delivery will have a significant effect on the engine. Refer to Routine maintenance for further information.
● Oil tank empty or low. This will have disastrous consequences if left unnoticed. Check and replenish tank regularly.
● Transmission oil low or worn out. Check the level regularly and investigate any loss of oil. If the oil level drops with no sign of external leakage it is likely that the crankshaft main bearing oil seals are worn, allowing transmission oil to be drawn into the crankcase during induction.

20 Miscellaneous causes

● Engine fins clogged. A build-up of mud in the cylinder head and cylinder barrel cooling fins will decrease the cooling capabilities of the fins. Clean the fins as required.

Clutch operating problems

21 Clutch slip

● No clutch lever play. Adjust clutch lever end play according to the procedure in Routine maintenance.
● Friction plates worn or warped. Overhaul clutch assembly, replacing plates out of specification.
● Plain plates worn or warped. Overhaul clutch assembly, replacing plates out of specification.
● Clutch spring(s) broken or worn. Old or heat-damaged (from slipping clutch) spring(s) should be renewed.
● Clutch lifting mechanism not adjusted properly. See Routine maintenance.
● Clutch inner cable snagging. Caused by a frayed cable or kinked outer cable. Replace the cable with a new one. Repair of a frayed cable is not advised.
● Clutch lifting mechanism defective. Worn or damaged parts in the clutch lifting mechanism could include the actuating arm or worm gear. Renew parts as necessary.
● Clutch centre and outer drum or body and gear (as appropriate) worn. Severe indentation by the clutch plate tangs will cause snagging of the plates preventing correct engagement. If this damage occurs, renewal of the worn components is required.
● Lubricant incorrect. Use of a transmission lubricant other than that specified may allow the plates to slip.

22 Clutch drag

● Clutch lever play excessive. Adjust lever at bars or at cable end if necessary.
● Clutch plates warped or damaged. This will cause a drag on the clutch, causing the machine to creep. Overhaul clutch assembly.

● Clutch spring tension uneven. Usually caused by a sagged or broken spring. Check and renew spring(s).
● Transmission oil deteriorated. Badly contaminated transmission oil and a heavy deposit of oil sludge on the plates will cause plate sticking. The oil recommended for this machine is of the detergent type, therefore it is unlikely that this problem will arise unless regular oil changes are neglected.
● Transmission oil viscosity too high. Drag in the plates will result from the use of an oil with too high a viscosity. In very cold weather clutch drag may occur until the engine has reached operating temperature.
● Clutch centre and outer drum or body and gear (as appropriate) worn. Indentation by the clutch plate tangs will prevent easy plate disengagement. If the damage is light the affected areas may be dressed with a fine file. More pronounced damage will necessitate renewal of the components.
● Clutch outer drum or gear (as appropriate) seized to shaft. Lack of lubrication, severe wear or damage can cause the outer drum or gear to seize to the shaft. Overhaul of the clutch, and perhaps the transmission, may be necessary to repair damage.
● Clutch lifting mechanism defective. Worn or damaged lifting mechanism parts can stick and fail to provide leverage. Overhaul clutch cover components.
● Loose clutch centre nut (125 and 150 models). Causes drum and centre misalignment, putting a drag on the engine. Engagement adjustment continually varies. Overhaul clutch assembly.

Gear selection problems

23 Gear lever does not return

● Weak or broken centraliser spring. Renew the spring.
● Gearchange shaft bent or seized. Distortion of the gearchange shaft often occurs if the machine is dropped heavily on the gear lever. Provided that damage is not severe straightening of the shaft is permissible.

24 Gear selection difficult or impossible

● Clutch not disengaging fully. See Section 22.
● Gearchange shaft bent. This often occurs if the machine is dropped heavily on the gear lever. Straightening of the shaft is permissible if the damage is not too great.
● Gearchange arms, pawls or pins worn or damaged. Wear or breakage of any of these items may cause difficulty in selecting one or more gears. Overhaul the selector mechanism.
● Gearchange drum stopper cam or detent arm and plunger damaged. Failure, rather than wear of these items may jam the drum thereby preventing gearchanging or causing false selection at high speed.
● Selector forks bent or seized. This can be caused by dropping the machine heavily on the gearchange lever or as a result of lack of lubrication. Though rare, bending of a shaft can result from a missed gearchange or false selection at high speed.
● Selector fork end and pin wear. Pronounced wear of these items and the grooves in the gearchange drum can lead to imprecise selection and, eventually, no selection. Renewal of the worn components will be required.
● Structural failure. Failure of any one component of the selector rod and change mechanism will result in improper or fouled gear selection.

25 Jumping out of gear

● Detent arm or plunger assembly worn or damaged. Wear of the arm or plunger and the cam with which it locates and breakage of the detent spring can cause imprecise gear selection resulting in jumping out of gear. Renew the damaged components.
● Gear pinion dogs worn or damaged. Rounding off the dog edges and the mating recesses in adjacent pinion can lead to jumping out of gear

when under load. The gears should be inspected and renewed. Attempting to reprofile the dogs is not recommended.
● Selector forks, gearchange drum and pinion grooves worn. Extreme wear of these interconnected items can occur after high mileages especially when lubrication has been neglected. The worn components must be renewed.
● Gear pinions, bushes and shafts worn. Renew the worn components.
● Bent gearchange shaft. Often caused by dropping the machine on the gear lever.
● Gear pinion tooth broken. Chipped teeth are unlikely to cause jumping out of gear once the gear has been selected fully; a tooth which is completely broken off, however, may cause problems in this respect and in any event will cause transmission noise.

26 Overselection

● Detent arm or plunger worn or broken. Renew the damaged items.
● Stopper arm spring worn or broken. Renew the spring.
● Gearchange arm stop pads worn. Repairs can be made by welding and reprofiling with a file.
● Selector limiter claw components (where fitted) worn or damaged. Renew the damaged items.

Abnormal engine noise

27 Knocking or pinking

● See Section 16.

28 Piston slap or rattling from cylinder

● Cylinder bore/piston clearance excessive. Resulting from wear, or partial seizure. This condition can often be heard as a high, rapid tapping noise when the engine is under little or no load, particularly when power is just beginning to be applied. Reboring to the next correct oversize should be carried out and a new oversize piston fitted.
● Connecting rod bent. This can be caused by over-revving, trying to start a very badly flooded engine (resulting in a hydraulic lock in the cylinder) or by earlier mechanical failure. Attempts at straightening a bent connecting rod are not recommended. Careful inspection of the crankshaft should be made before renewing the damaged connecting rod.
● Gudgeon pin, piston boss bore or small-end bearing wear or seizure. Excess clearance or partial seizure between normal moving parts of these items can cause continuous or intermittent tapping noises. Rapid wear or seizure is caused by lubrication starvation.
● Piston rings worn, broken or sticking. Renew the rings after careful inspection of the piston and bore.

29 Other noises

● Big-end bearing wear. A pronounced knock from within the crankcase which worsens rapidly is indicative of big-end bearing failure as a result of extreme normal wear or lubrication failure. Remedial action in the form of a bottom end overhaul should be taken; continuing to run the engine will lead to further damage including the possibility of connecting rod breakage.
● Main bearing failure. Extreme normal wear or failure of the main bearings is characteristically accompanied by a rumble from the crankcase and vibration felt through the frame and footrests. Renew the worn bearings and carry out a very careful examination of the crankshaft.
● Crankshaft excessively out of true. A bent crank may result from over-revving or damage from an upper cylinder component or gearbox failure. Damage can also result from dropping the machine on either

crankshaft end. Straightening of the crankshaft is not be possible in normal circumstances; a replacement item should be fitted.

● Engine mounting loose. Tighten all the engine mounting nuts and bolts.

● Cylinder head gasket leaking. The noise most often associated with a leaking head gasket is a high pitched squeaking, although any other noise consistent with gas being forced out under pressure from a small orifice can also be emitted. Gasket leakage is often accompanied by oil seepage from around the mating joint or from the cylinder head holding down bolts and nuts. Leakage results from insufficient or uneven tightening of the cylinder head fasteners, or from random mechanical failure. Retightening to the correct torque figure will, at best, only provide a temporary cure. The gasket should be renewed at the earliest opportunity.

● Exhaust system leakage. Popping or crackling in the exhaust system, particularly when it occurs with the engine on the overrun, indicates a poor joint either at the cylinder port or at the exhaust pipe/silencer connection. Looseness of the clamp should be looked for.

Abnormal transmission noise

30 Clutch noise

● Clutch outer drum/friction plate tang clearance excessive (125 and 150 models).

● Primary drive gear teeth worn or damaged (250, 251 and 300 models).

● Primary drive gear endfloat excessive (250, 251 and 300 models).

● Primary drive chain freeplay excessive (125 and 150 models).

● Clutch drum/primary drive sprocket not correctly aligned (125 and 150 models).

31 Transmission noise

● Bearing or bushes worn or damaged. Renew the affected components.

● Gear pinions worn or chipped. Renew the gear pinions.

● Metal chips jammed in gear teeth.This can occur when pieces of metal from any failed component are picked up by a meshing pinion. The condition will lead to rapid bearing wear or early gear failure.

● Transmission oil level too low. Top up immediately to prevent damage to gearbox and engine.

● Gearchange mechanism worn or damaged. Wear or failure of certain items in the selection and change components can induce mis-selection of gears (see Section 24) where incipient engagement of more than one gear set is promoted. Remedial action, by the overhaul of the gearbox, should be taken without delay.

● Chain snagging on cases or cycle parts. A badly worn chain or one that is excessively loose may snag or smack against adjacent components.

Exhaust smokes excessively

32 White/blue smoke (caused by oil burning)

● Petrol/oil ratio incorrect. Ensure that oil is mixed with the petrol in the correct ratio. The manufacturer's recommendation must be adhered to if excessive smoking or under-lubrication is to be avoided.

● Oil pump settings incorrect. Check and reset the oil pump as described in Routine maintenance.

● Crankshaft main bearing oil seals worn. Wear in the main bearing oil seals, often in conjunction with wear in the bearings themselves, can allow transmission oil to find its way into the crankcase and thence to the combustion chamber. This condition is often indicated by a mysterious drop in the transmission oil level with no sign of external leakage.

● Accumulated oil deposits in exhaust system. If the machine is used for short journeys only it is possible for the oil residue in the exhaust gases to condense in the relatively cool silencer. If the machine is then

taken for a longer run in hot weather, the accumulated oil will burn off producing ominous smoke from the exhaust.

33 Black smoke (caused by over-rich mixture)

● Air filter element clogged. Clean or renew the element.

● Main jet loose or too large. Remove the float chamber to check for tightness of the jet. If the machine is used at high altitudes rejetting will be required to compensate for the lower atmospheric pressure.

● Cold start mechanism jammed on. Check that the mechanism works smoothly and correctly and that the operating cable is lubricated and not snagged.

● Fuel level too high. The fuel level is controlled by the float height which can increase as a result of wear or damage. Remove the float bowl and check the float height. Check also that floats have not punctured; a punctured float will lose buoyancy and allow an increased fuel level.

● Float valve needle stuck open. Caused by dirt or a worn valve. Clean the float chamber or renew the needle and, if necessary, the valve seat.

Poor handling or roadholding

34 Directional instability

● Steering head bearings worn or damaged. Correct adjustment of the bearing will prove impossible to achieve if wear or damage has occurred. Inconsistent handling will occur including rolling or weaving at low speed and poor directional control at indeterminate higher speeds. The steering head bearing should be dismantled for inspection and renewed if required. Lubrication should also be carried out.

● Bearing races pitted or dented. Impact damage caused, perhaps, by an accident or riding over a pot-hole can cause indentation of the bearing, usually in one position. This should be noted as notchiness when the handlebars are turned. Renew and lubricate the bearings.

● Steering stem bent. This will occur only if the machine is subjected to a high impact such as hitting a curb or a pot-hole. The lower yoke/stem should be renewed; do not attempt to straighten the stem.

● Front or rear tyre pressures too low.

● Front or rear tyre worn. General instability, high speed wobbles and skipping over white lines indicates that tyre renewal may be required. Tyre induced problems, in some machine /tyre combinations, can occur even when the tyre in question is by no means fully worn.

● Swinging arm bearings worn. Difficulties in holding line, particularly when cornering or when changing power settings indicates wear in the swinging arm bearings. The swinging arm should be removed from the machine and the bearings renewed.

● Swinging arm flexing. The symptoms given in the preceding paragraph will also occur if the swinging arm fork flexes badly. This can be caused by structural weakness as a result of corrosion, fatigue or impact damage, or because the rear wheel spindle is slack.

● Wheel bearings worn. Renew the worn bearings.

● Loose wheel spokes. The spokes should be tightened evenly to maintain tension and trueness of the rim.

● Tyres unsuitable for machine. Not all available tyres will suit the characteristics of the frame and suspension, indeed, some tyres or tyre combinations may cause a transformation in the handling characteristics. If handling problems occur immediately after changing to a new tyre type or make, revert to the original tyres to see whether an improvement can be noted. In some instances a change to what are, in fact, suitable tyres may give rise to handling deficiencies. In this case a thorough check should be made of all frame and suspension items which affect stability.

35 Steering bias to left or right

● Rear wheel out of alignment. Caused by uneven adjustment of chain tensioner adjusters allowing the wheel to be askew in the fork ends. A bent rear wheel spindle will also misalign the wheel in the swinging arm.

● Wheels out of alignment. This can be caused by impact damage to the frame, swinging arm, wheel spindles or front forks. Although occasionally a result of material failure or corrosion it is usually as a result of a crash.

● Front forks twisted in the steering yokes. A light impact, for instance with a pot-hole or low curb, can twist the fork legs in the steering yokes without causing structural damage to the fork legs or the yokes themselves. Re-alignment can be made by loosening the yoke pinch bolts, wheel spindle and mudguard bolts. Re-align the wheel with the handlebars and tighten the bolts working upwards from the wheel spindle. This action should be carried out only when there is no chance that structural damage has occurred.

36   Handlebar vibrates or oscillates

● Tyres worn or out of balance. Either condition, particularly in the front tyre, will promote shaking of the fork assembly and thus the handlebars. A sudden onset of shaking can result if a balance weight is displaced during use.

● Tyres badly positioned on the wheel rims. A moulded line on each wall of a tyre is provided to allow visual verification that the tyre is correctly positioned on the rim. A check can be made by rotating the tyre; any misalignment will be immediately obvious.

● Wheel rims warped or damaged. Inspect the wheels for runout as described in Routine maintenance.

● Swinging arm bushes worn. Renew the bushes.

● Wheel bearings worn. Renew the bearings.

● Loose fork component fasteners. Loose nuts and bolts holding the fork legs, wheel spindle, mudguards or steering stem can promote shaking at the handlebars. Fasteners on running gear such as the forks and suspension should be check tightened occasionally to prevent dangerous looseness of components occurring.

● Engine mounting bolts loose. Tighten all fasteners.

37   Poor front fork performance

● Damping fluid level incorrect. If the fluid level is too low poor suspension control will occur resulting in a general impairment of roadholding and early loss of tyre adhesion when cornering and braking. Too much oil is unlikely to change the fork characteristics unless severe overfilling occurs when the fork action will become stiffer and oil seal failure may occur.

● Damping oil viscosity incorrect. The damping action of the fork is directly related to the viscosity of the damping oil. The lighter the oil used, the less will be the damping action imparted. For general use, use the recommended viscosity of oil, changing to a slightly higher or heavier oil only when a change in damping characteristic is required. Overworked oil, or oil contaminated with water which has found its way past the seals, should be renewed to restore the correct damping performance and to prevent bottoming of the forks.

● Damping components worn or corroded. Advanced normal wear of the fork internals is unlikely to occur until a very high mileage has been covered. Continual use of the machine with damaged oil seals which allows the ingress of water, or neglect, will lead to rapid corrosion and wear. Dismantle the forks for inspection and overhaul.

● Weak fork springs. Progressive fatigue of the fork springs, resulting in a reduced spring free length, will occur after extensive use. This condition will promote excessive fork dive under braking, and in its advanced form will reduce the at-rest extended length of the forks and thus the fork geometry. Renewal of the springs as a pair is the only satisfactory course of action.

● Bent stanchions or corroded stanchions. Both conditions will prevent correct telescoping of the fork legs, and in an advanced state can cause sticking of the fork in one position. In a mild form corrosion will cause stiction of the fork thereby increasing the time the suspension takes to react to an uneven road surface. Bent fork stanchions should be attended to immediately because they indicate that impact damage has

occurred, and there is a danger that the forks will fail with disastrous consequences.

38   Front fork judder when braking (see also Section 50)

● Wear between the fork stanchions and the fork legs. Renewal of the affected components is required.

● Worn steering head bearings. Renew the bearings.

● Warped brake disc or drum. If irregular braking action occurs fork judder can be induced in what are normally serviceable forks. Renew the damaged brake components.

39   Poor rear suspension performance

● Rear suspension unit damper worn out or leaking. The damping performance of most rear suspension units falls off with age. This is a gradual process, and thus may not be immediately obvious. Indications of poor damping include hopping of the rear end when cornering or braking, and a general loss of positive stability.

● Weak rear springs. If the suspension unit springs fatigue they will promote excessive pitching of the machine and reduce the ground clearance when cornering. Although replacement springs are available separately from the rear suspension damper unit it is probable that if spring fatigue has occurred the damper units will also require renewal.

● Swinging arm flexing or bushes worn. See Sections 34 and 36.

● Bent suspension unit damper rod. This is likely to occur only if the machine is dropped or if seizure of the piston occurs. If either happens the suspension units should be renewed as a pair.

Abnormal frame and suspension noise

40   Front end noise

● Oil level low or too thin. This can cause a 'spurting' sound and is usually accompanied by irregular fork action.

● Spring weak or broken. Makes a clicking or scraping sound. Fork oil will have a lot of metal particles in it.

● Steering head bearings worn. Clicks when braking. Check and renew.

● Fork stanchion loose. Make sure the bottom yoke pinch bolts and fork top bolts are tight.

● Fork stanchion bent. Good possibility if machine has been dropped. Repair or renew stanchion.

41   Rear suspension noise

● Fluid level too low. Leakage of a suspension unit, usually evident by oil on the outer surfaces, can cause a spurting noise. The suspension units should be overhauled as a pair.

● Defective rear suspension unit with internal damage. Renew the suspension units as a pair.

Brake problems

42   Brakes are spongy or ineffective – disc brakes

● Air in brake circuit. This is only likely to happen in service due to neglect in checking the fluid level or because a leak has developed. The problem should be identified and the brake system bled of air.

● Pads worn. Check the pad wear as described in Routine maintenance and renew the pads if necessary.

● Contaminated pads. Cleaning pads which have been contaminated with oil, grease or brake fluid is unlikely to prove successful; the pads should be renewed.

● Pads glazed. This is usually caused by overheating. The surface of the pads may be roughened using glass-paper or a fine file.

● Brake fluid deterioration. A brake which on initial operation is firm but rapidly becomes spongy in use may be failing due to water contamination of the fluid. The fluid should be drained and then the system refilled and bled.

● Master cylinder seal failure. Wear or damage of master cylinder internal parts will prevent pressurisation of the brake fluid. Overhaul the master cylinder unit.

● Caliper seal failure. This will almost certainly be obvious by loss of fluid, a lowering of fluid in the master cylinder reservoir and contamination of the brake pads and caliper. Overhaul the caliper assembly.

● Brake lever or pedal improperly adjusted. Adjust the clearance between the lever end and master cylinder plunger to take up lost motion, as recommended in Routine maintenance.

43 Brakes drag – disc brakes

● Disc warped. The disc must be renewed.

● Caliper piston, caliper or pads corroded. The brake caliper assembly is vulnerable to corrosion due to water and dirt, and are cleaned at regular intervals and lubricated in the recommended manner, will become sticky in operation.

● Piston seal deteriorated. The seal is designed to return the piston in the caliper to the retracted position when the brake is released. Wear or old age can affect this function. The caliper should be overhauled if this occurs.

● Brake pad damaged. Pad material separating from the backing plate due to wear or faulty manufacture. Renew the pads. Faulty installation of a pad also will cause dragging.

● Wheel spindle bent. The spindle may be straightened if no structural damage has occurred.

● Brake lever or pedal not returning. Check that the lever or pedal works smoothly throughout its operating range and does not snag on any adjacent cycle parts. Lubricate the pivot if necessary.

44 Brake lever or pedal pulsates in operation – disc brakes

● Disc warped or irregularly worn. The disc must be renewed.

● Wheel spindle bent. The spindle may be straightened provided no structural damage has occurred.

45 Disc brake noise

● Brake squeal. Squealing can be caused by dust on the pads, usually in combination with glazed pads, or other contamination from oil, grease, brake fluid or corrosion. Persistent squealing which cannot be traced to any of the normal causes can often be cured by applying a thin layer of high temperature silicone grease to the rear of the pads. Make absolutely certain that no grease is allowed to contaminate the braking surface of the pads.

● Glazed pads. This is usually caused by high temperatures or contamination. The pad surfaces may be roughened using glass-paper or a fine file. If this approach does not effect a cure the pads should be renewed.

● Disc warped. This can cause a chattering, clicking or intermittent squeal and is usually accompanied by a pulsating brake lever or pedal or uneven braking. The disc must be renewed.

● Brake pads fitted incorrectly or undersize. Longitudinal play in the pads due to omission of the locating springs (where fitted) or because pads of the wrong size have been fitted will cause a single tapping noise every time the brake is operated. Inspect the pads for correct installation and security.

46 Brakes are spongy or ineffective – drum brakes

● Brake cable deterioration. Damage to the outer cable by stretching or being trapped will give a spongy feel to the brake lever. The cable should be renewed. A cable which has become corroded due to old age or neglect of lubrication will partially seize making operation very heavy.

Lubrication at this stage may overcome the problem but the fitting of a new cable is recommended.

● Worn brake linings. Determine lining wear by removing the wheel and withdrawing the brake backplate. Renew the shoes as a pair if the linings are worn below the recommended limit.

● Worn brake camshaft. Wear between the camshaft and the bearing surface will reduce brake feel and reduce operating efficiency. Renewal of one or both items will be required to rectify the fault.

● Worn brake cam and shoe ends. Renew the worn components.

● Linings contaminated with dust or grease. Any accumulations of dust should be cleaned from the brake assembly and drum using a petrol dampened cloth. Do not blow or brush off the dust because it is asbestos based and thus harmful if inhaled. Light contamination from grease can be removed from the surface of the brake linings using a solvent; attempts at removing heavier contamination are less likely to be successful because some of the lubricant will have been absorbed by the lining material which will severely reduce the braking performance.

47 Brake drag – drum brakes

● Incorrect adjustment. Re-adjust the brake operating mechanism.

● Drum warped or oval. This can result from overheating or impact or uneven tension of the wheel spokes. The condition is difficult to correct, although if slight ovality only occurs, skimming the surface of the brake drum can provide a cure. This is work for a specialist engineer. Renewal of the complete wheel hub is normally the only satisfactory solution.

● Weak brake shoe return springs. This will prevent the brake lining/shoe units from pulling away from the drum surface once the brake is released. The springs should be renewed.

● Brake camshaft, lever pivot or cable poorly lubricated. Failure to attend to regular lubrication of these areas will increase operating resistance which, when compounded, may cause tardy operation and poor release movement.

48 Brake lever or pedal pulsates in operation – drum brakes

● Drums warped or oval. This can result from overheating or impact or uneven spoke tension. This condition is difficult to correct, although if slight ovality only occurs skimming the surface of the drum can provide a cure. This is work for a specialist engineer. Renewal of the hub is normally the only satisfactory solution.

49 Drum brake noise

● Drum warped or oval. This can cause intermittent rubbing of the brake linings against the drum. See the preceding Section.

● Brake linings glazed. This condition, usually accompanied by heavy lining dust contamination, often induces brake squeal. The surface of the linings may be roughened using glass-paper or a fine file.

50 Brake induced fork judder

● Worn front fork stanchions and legs, or worn or badly adjusted steering head bearings. These conditions, combined with uneven or pulsating braking as described in Sections 44 and 48 will induce more or less judder when the brakes are applied, dependent on the degree of wear and poor brake operation. Attention should be given to both areas of malfunction. See the relevant Sections.

Electrical problems

51 Battery dead or weak

● Battery faulty. Battery life should not be expected to exceed 3 to 4 years. Gradual sulphation of the plates and sediment deposits will reduce the battery performance. Plate and insulator damage can often occur as a result of vibration. Complete power failure, or intermittent failure, may be due to a broken battery terminal. Lack of electrolyte will prevent the battery maintaining charge.

● Battery leads making poor contact. Remove the battery leads and clean them and the terminals, removing all traces of corrosion and tarnish. Reconnect the leads and apply a coating of petroleum jelly to the terminals.

● Load excessive. If additional items such as spot lamps, are fitted, which increase the total electrical load above the maximum alternator output, the battery will fail to maintain full charge. Reduce the electrical load to suit the electrical capacity.

● Rectifier failure.

● Alternator rotor or stator coils open-circuit or shorted.

● Charging circuit shorting or open circuit. This may be caused by frayed or broken wiring, dirty connectors or a faulty ignition switch. The system should be tested in a logical manner. See Section 54.

52   Battery overcharged

● Regulator faulty. Overcharging is indicated if the battery becomes hot or it is noticed that the electrolyte level falls repeatedly between checks. In extreme cases the battery will boil causing corrosive gases and electrolyte to be emitted through the vent pipes.

● Battery wrongly matched to the electrical circuit. Ensure that the specified battery is fitted to the machine.

53   Total electrical failure

● Fuse blown. Check the main fuse. If a fault has occurred, it must be rectified before a new fuse is fitted.

● Battery faulty. See Section 51.

● Earth failure. Check that the frame main earth strap from the battery is securely affixed to the frame and is making a good contact.

● Ignition switch or power circuit failure. Check for current flow through the battery positive lead (red) to the ignition switch. Check the ignition switch for continuity.

54   Circuit failure

● Cable failure. Refer to the machine's wiring diagram and check the circuit for continuity. Open circuits are a result of loose or corroded connections, either at terminals or in-line connectors, or because of broken wires. Occasionally, the core of a wire will break without there being any apparent damage to the outer plastic cover.

● Switch failure. All switches may be checked for continuity in each switch position, after referring to the switch position boxes incorporated in the wiring diagram for the machine. Switch failure may be a result of mechanical breakage, corrosion or water.

● Fuse blown. Refer to the wiring diagram to check whether or not a circuit fuse is fitted. Replace the fuse, if blown, only after the fault has been identified and rectified.

55   Bulbs blowing repeatedly

● Vibration failure. This is often an inherent fault related to the natural vibration characteristics of the engine and frame and is, thus, difficult to resolve. Modifications of the lamp mounting, to change the damping characteristics, may help.

● Intermittent earth. Repeated failure of one bulb, particularly where the bulb is fed directly from the generator, indicates that a poor earth exists somewhere in the circuit. Check that a good contact is available at each earthing point in the circuit.

● Reduced voltage. Where a quartz-halogen bulb is fitted the voltage to the bulb should be maintained or early failure of the bulb will occur. Do not overload the system with additional electrical equipment in excess of the system's power capacity and ensure that all circuit connections are maintained clean and tight.

Routine maintenance

Refer to Chapter 7 for information relating to 1991-on models

Specifications

Engine

	Isolator	NGK
Spark plug:		
250 and 251 models	ZM 14-260	B7HS
All other models	ZM 14-260	B8HS
Electrode gap	0.6 mm (0.024 in)	
Ignition timing:		
Standard	2.5 mm (0.10 in) BTDC	
Tolerance:		
300 models	2.5 – 2.7 mm (0.10 – 0.11 in)	
All other models	2.5 – 3.0 mm (0.10 – 0.12 in)	
Contact breaker gap:		
Standard	0.3 mm (0.012 in)	
Tolerance	0.3 – 0.4 mm (0.012 – 0.016 in)	
Idle speed	Approximately 1200 rpm	
Clutch cable free play – at handlebar lever butt end	3 mm (0.12 in)	

Cycle parts

	Front	Rear
Front brake lever free play (drum brake models) – at handlebar lever butt end	3 – 5 mm (0.12 – 0.20 in)	
Brake pad friction material service limit	0.5 mm (0.020 in)	
Tyre pressures – tyres cold:	**Front**	**Rear**
Solo:		
251 models	1.7 kg/cm^2 (24 psi)	1.9 kg/cm^2 (27 psi)
All other models	1.5 kg/cm^2 (21 psi)	1.9 kg/cm^2 (27 psi)
Up to maximum permissible load*:		
125 and 150 models	1.5 kg/cm^2 (21 psi)	2.7 kg/cm^2 (38 psi)
250 and 300 models	1.5 kg/cm^2 (21 psi)	2.5 kg/cm^2 (35 psi)
251 models	1.7 kg/cm^2 (24 psi)	2.7 kg/cm^2 (38 psi)

* Note: maximum permissible load comprises the weight of the machine, rider, passenger and luggage carried. On 125 and 150 models this is 290 kg (639 lb), and on 250, 251 and 300 models this is 330 kg (726 lb).

Recommended lubricants and fluids

Fuel grade	Unleaded or low-lead, minimum octane rating 88 RON/RM
Engine lubrication – pre-mix models:	
Recommended oil	Good quality self-mixing two-stroke oil
Fuel/oil pre-mix ratio	50:1 (0.16 pint/136 cc oil to 1 gal petrol, 30 cc oil to 1 litre petrol)
Engine lubrication – oil injection models:	
Recommended oil	Good quality branded two-stroke oil
Oil tank capacity	Approximately 1.3 litres (0.27 Imp gal)
Transmission lubrication:	
Capacity:	
125 and 150 models	500 cc (0.88 Imp pint)
250, 251 and 300 models	900 cc (1.58 Imp pint)
Recommended oil – all models	Good quality EP80 hypoid gear oil
Front forks – capacity per leg:	
Normal	230 cc (8.09 fl oz)
Maximum:	
125 and 150 models	250 cc (8.79 fl oz)
250, 251 and 300 models	265 cc (9.32 fl oz)
Fork oil level – from bottom of fork:	
125 and 150 models:	
Normal	350 mm (13.78 in)
Maximum	370 mm (14.57 in)
250, 251 and 300 models:	
Normal	330 mm (12.99 in)
Maximum	395 mm (15.55 in)
Fork oil – all models	SAE10 fork oil
Drive chain lubrication	Aerosol chain lubricant or gear oil
Brake fluid (disc brake models)	DOT 3
Steering head and wheel bearings, and swinging arm pivot shaft	Good quality high melting-point grease
Instrument drive cables and centre stand pivot	General purpose grease
Control cables and all other pivots	Engine oil or light machine oil

Introduction

Periodic routine maintenance is a continuous process which should commence immediately the machine is used. The object is to maintain all adjustments and to diagnose and rectify minor defects before they develop into more extensive, and often more expensive, problems.

It follows that if the machine is maintained properly, it will both run and perform with optimum efficiency, and be less prone to unexpected breakdowns. Regular inspection of the machine will show up any parts which are wearing, and with a little experience, it is possible to obtain the maximum life from any one component, renewing it when it becomes so worn that it is liable to fail.

Regular cleaning can be considered as important as mechanical maintenance. This will ensure that all the cycle parts are inspected regularly and are kept free from accumulations of road dirt and grime. Cleaning is especially important during the winter months, despite its appearance of being a thankless task which very soon seems pointless. On the contrary, it is during these months that the paintwork, chromium plating, and the alloy casings suffer the ravages of abrasive grit, rain and road salt. A couple of hours spent weekly on cleaning the machine will maintain its appearance and value, and highlight small points, such as chipped paint, before they become a serious problem.

It should be noted that the intervals between each maintenance task serve only as a guide. As the machine gets older, or if it is used under particularly arduous conditions, it is advisable to reduce the period between each check.

For ease of reference, most service operations are described in detail under the relevant heading. However, if further general information is required, this can be found under the pertinent section heading and chapter in the main text.

Although no special tools are required for routine maintenance, a good selection of general workshop tools is essential; refer to the Maintenance and minor repair toolkit section of Tools and working facilities.

Daily (pre-ride) checks

It is recommended that the following items are checked whenever the machine is about to be used. This is important to prevent the risk of unexpected failure of any component while riding the machine and, with experience, can be reduced to a simple checklist which will only take a few moments to complete. For those owners who are not inclined to check all items with such frequency, it is suggested that the best course is to carry out the checks in the form of a service which can be undertaken each week or before any long journey. It is essential that all items are checked and serviced with reasonable frequency.

1 Check the engine oil level (oil injection models only)

Before setting out on any journey it is essential to check the level of oil in the tank, mounted below the left-hand sidepanel. Although the tank is fitted with a sightglass, a more accurate check can be made using the dipstick which is fitted to the filler cap. Unscrew the filler cap, check the oil level using the filler cap dipstick, and top up as necessary. Although not strictly necessary, it is strongly recommended that the oil level be kept topped up to the base of the filler neck at all times. Refit the filler cap and tighten it securely.

Note: In the event of the oil level falling low enough for air to enter the pipe between the oil tank and pump, the system must be bled before the machine is ridden. The bleeding operation is described in Chapter 2.

2 Check the fuel level

Checking the petrol level may seem obvious, but it is all too easy to forget. Ensure that you have enough petrol to complete your journey, or at least to get you to the nearest petrol station.

3 Check the battery

The battery is located behind the right-hand sidepanel which must first be removed to gain access to it.

Whenever the battery is disconnected, remember to disconnect the negative (–) terminal first, to prevent the possibility of short circuits. The electrolyte level, visible through the translucent casing, must be between the two level marks on the casing. If necessary, remove the cell caps and top up to the upper level using only distilled water. Check that the terminals are clean and apply a thin smear of petroleum jelly (not grease) to each to prevent corrosion.

On refitting, check that the vent hose is clear and that it is correctly routed with no kinks, also that it hangs well below any other component. Secure the battery with its bracket and tighten its retaining bolt securely. Always connect the negative (–) terminal last when refitting the battery. Check that the terminals are tight and that the fuse connections are clean and tight. Check also that all fuses are of the correct rating and that a spare is available on the machine.

At regular intervals remove the battery and check that there is no pale grey sediment deposited at the bottom of the casing. This is caused by sulphation of the plates as a result of a re-charge at too high a rate or as a result of the battery being left discharged for long periods. A good battery should have little or no sediment visible and its plates should be straight and pale brown or grey in colour. If the sediment deposits are deep enough to reach the bottom of the plates, or if the plates are buckled and have whitish deposits on them, the battery is faulty and must be renewed. Remember that a poor battery will give rise to a large number of electrical faults.

If the machine is not in regular use, disconnect the battery and give it a refresher charge every month to six weeks, as described in Chapter 6.

4 Check and adjust the brakes

Check that the front and rear brakes work effectively and without binding, then check them as follows.

Front brake

Note: *brake fluid will discolour or remove paint if contact is allowed. Avoid this where possible and remove accidental spillage immediately. Similarly, avoid contact with plastic components.*

On models fitted with a front disc brake, check that the fluid level in the master cylinder reservoir is between the maximum and minimum level marks. Both the maximum and minimum marks are cast on the reservoir body and the fluid level should be visible through the translucent plastic of the reservoir body.

If necessary, unscrew the cap, remove the ventilation ring and diaphragm and top up the fluid to the maximum level mark. Use only new brake fluid, of the specified type, from a sealed container to top up the reservoir. Ensure the diaphragm is clean, then refit it with the ventilation ring and screw the reservoir cap on securely.

Once the fluid level is known to be correct, it is necessary to check the operation of the stoplamp switch, located in the handlebar lever stock. The stoplamp should illuminate as soon as the brake starts to function. If not, slacken the stoplamp switch locknut and pull the front brake lever in until the pads are just in contact with the disc and the brake is just starting to grip. With the lever held steady in this position, unscrew the stoplamp switch so that the stoplamp goes out (if necessary), then slowly screw it in until the stoplamp comes on. Hold the stoplamp switch body and tighten its locknut securely.

On models with a front drum brake, place the machine on its centre stand and place a suitable stand beneath the crankcase so that the front wheel is raised off the ground. Using the handlebar mounted adjuster, adjust the cable so that there is 3 – 5 mm (0.08 – 0.20 in) of freeplay between the lever butt end and its mounting bracket before the brake shoes come in contact with the drum. If there is insufficient range in the adjuster the shoes must be removed for inspection as described under the three-monthly heading. Finally, check that the stoplamp switch is set correctly so that the lamp illuminates as soon as the shoes contact the drum. If necessary, adjust the switch position as described above.

Rear brake

Place the machine on its centre stand so that the rear wheel is raised clear of the ground. Adjust the rear brake, by means of the adjusting nut on the end of the brake rod, so that when you are seated in a normal riding position the brake pedal can be operated comfortably and full pressure can be applied. Although the manufacturer does not state a specific amount of freeplay for the rear brake, as a guide, it is recommended that there should be around 20 mm (0.8 in) of travel, measured at the lever tip, before the shoes come in contact with the drum.

Once the rear brake is correctly adjusted, check the operation of the stoplamp switch. The stoplamp should illuminate as soon as the brake is

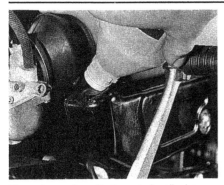
Use only good quality two-stroke oil when topping up oil tank

Electrolyte level must be between the marks on battery casing

Ensure fluid level is kept between marks on the front of master cylinder reservoir

Slide back rubber cover to reveal front stoplamp switch – adjust as described in text

On 251 models rear stoplamp switch is adjusted using plastic sleeve nut ...

... on all other models switch is adjusted by slackening the locknut and rotating the screw

beginning to engage, if not the switch must be adjusted as follows.

On 251 models the stoplamp switch is situated behind the right-hand sidepanel. Hold the body of the switch steady and rotate the plastic sleeve nut to raise or lower the switch as necessary; do not allow the body of the switch to rotate or its terminals will be damaged.

On all other models the switch is mounted in the rear brake backplate. Slacken the switch locknut and apply the rear brake pedal until the shows begin to contact the drum. With the pedal held in this position, slowly rotate the adjuster screw until the stoplamp goes out (if necessary), then rotate adjuster back until the stoplamp comes on. Hold the adjuster screw in this position and tighten its locknut securely.

5 Check the clutch

The clutch is correctly adjusted when there is 3 mm (0.12 in) of freeplay in the cable, measured between the handlebar lever butt end and its mounting clamp. If this is not the case, slacken the locknut on the handlebar mounted adjuster and rotate the adjuster until the required amount of freeplay is obtained. Tighten the locknut securely to complete adjustment. If there is insufficient range in the adjuster, screw it in to obtain the maximum cable freeplay possible and adjust the clutch using the procedure given under the relevant sub heading.

125 and 150 models

Remove the small circular cap from the right-hand crankcase cover to reveal the clutch pushrod adjuster. Slacken the adjuster locknut and screw in the adjuster screw until light pressure can be felt acting on the screw. Do not overtighten the adjuster as this is the point at which the clutch pressure plate is starting to lift the pressure plate. Once pressure is felt, back the adjuster screw off by ¾ of a turn and secure it by tightening its locknut securely. Refit the circular cap to the casing and readjust the cable using the handlebar-mounted adjuster as described above.

250, 251 and 300 models

Release the three screws which retain the tachometer drive housing or cover (as appropriate) to the left-hand crankcase cover and remove it from the engine unit. Displace the rubber dust cover from the clutch

cable mounting sleeve, situated on the front of the left-hand crankcase cover, then pull out the clutch cable and remove its retaining spacer. Unscrew the mounting sleeve from the crankcase cover and slide it along the clutch cable.

Keeping the clutch cable taut, rotate the slotted adjuster plate, situated behind the tachometer drive, until the distance from the edge of the crankcase cover to the centre of the clutch cable nipple is 11 mm (0.43 in). Once this distance is correct, refit the tachometer drive housing or cover (as appropriate) to the casing, ensuring that it engages correctly with its drive gear, and tighten its three retaining screws securely. Refit the clutch mounting sleeve to the casing, tightening it securely, then refit the clutch cable retaining spacer and the dust cover. Finally adjust the clutch cable using the handlebar-mounted adjuster as described above.

6 Check the tyres

Check the tyre pressures with a gauge that is known to be accurate. It is worthwhile purchasing a pocket gauge for this purpose as the gauges on garage forecourt airlines are notoriously inaccurate. The pressures should be checked with the tyres cold. Even a few miles travelled will warm up the tyres to a point where the pressures increase and an inaccurate reading will result.

At the same time as the tyre pressures are checked, examine the tyres themselves. Check them for damage, especially for splitting of the sidewalls. Remove any stones or other road debris caught between the treads. This is particularly important on the rear tyre, where rapid deflation due to penetration of the inner tube will almost certainly cause loss of control. When checking the tyres for damage, the depth of the tread should also be checked. For UK machines, it is vital to keep the tread depth within the legal limits of 1 mm of depth over three-quarters of the tread width around the entire circumference with no sign of bald patches. Many riders, however, consider nearer 2 mm to be the limit for secure roadholding, traction and braking in adverse weather conditions.

The tyres should be renewed whenever they are found to be damaged or excessively worn (see Chapter 5).

7 Check the controls

Check that the throttle, clutch and brake controls operate smoothly,

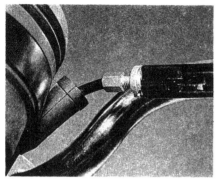

Clutch cable freeplay is adjusted using handlebar-mounted adjuster

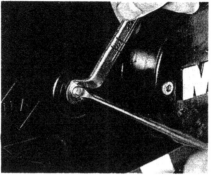

Adjusting clutch lifting mechanism – 125 and 150 models

On 250, 251 and 300 models remove tachometer drive housing or cover ...

... and remove clutch cable retaining spacer and mounting sleeve

Rotate slotted adjuster plate...

... until distance from centre of nipple to casing is 11 mm (0.43 in)

and ensure that the gear lever and footrests are securely fastened. If a bolt is going to work loose, or a cable snap, it is better that it is discovered at this stage with the machine at a standstill, rather than when it is being ridden. Check the operation of the throttle cable and ensure that it snaps shut immediately it is released. If any of the operating cables on the machine appear dry or stiff in operation, apply a few drops of light machine oil to their exposed sections. If this fails to cure the problem, the cable must be removed for thorough lubrication as described under the annual heading.

8 Legal check

Check that all lights, turn signals, horn and speedometer are working correctly to make sure the machine complies with all legal requirements in this respect. Check also that the headlamp is correctly aimed. The headlamp aim must be set with the rider (and pillion passenger, if one is regularly carried) seated normally on the machine. The dip beam centre (as shown on a wall 25 feet away) must be at the same height from the ground as the centre of the headlamp itself. The headlamp is adjusted by slackening the nut which secures it to the bottom yoke, moving the headlamp unit to the required position and then tightening its mounting nut securely.

Three monthly, or every 1550 miles (2500 km)

Perform all the tasks listed under the pre-ride checks, then carry out the following.

1 Check the spark plug

Since the manufacturer's specified make and type of spark plug is not readily available, a great deal of care is required when selecting plugs. Although it is difficult to give precise equivalents due to the difference in manufacturer's heat ranges, the nearest equivalents available in the UK are listed in the Specifications at the start of this Chapter.

If the spark plug is suspected of being faulty it can be tested only by the substitution of a brand new (not second-hand) plug of the correct make, type and heat range.

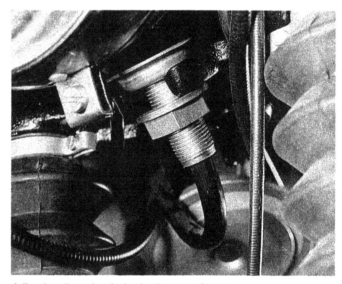

Adjust headlamp by slackening its mounting nut

Note that alternatives to the standard plug are available, although the advice of an authorized MZ dealer or similar expert should be sought before the plug heat range is altered from standard. The use of too cold, or hard, a grade of plug will result in fouling and the use of too hot, or soft, a grade of plug will result in engine damage due to excess heat being generated. If the correct grade is fitted, however, it will be possible to use the condition of the spark plug electrodes to diagnose a fault in the engine or to show if the engine is operating efficiently or not. The accompanying series of colour photographs will show this clearly.

It is advisable to carry a new spark plug on the machine, having first set the electrodes to the correct gap. Whilst spark plugs do not often fail, a new replacement is well worth having if a breakdown does occur. Ensure that the spare is of the correct heat range and type.

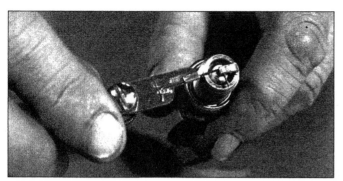

Electrode gap check - use a wire type gauge for best results

Electrode gap adjustment - bend the side electrode using the correct tool

Normal condition - A brown, tan or grey firing end indicates that the engine is in good condition and that the plug type is correct

Ash deposits - Light brown deposits encrusted on the electrodes and insulator, leading to misfire and hesitation. Caused by excessive amounts of oil in the combustion chamber or poor quality fuel/oil

Carbon fouling - Dry, black sooty deposits leading to misfire and weak spark. Caused by an over-rich fuel/air mixture, faulty choke operation or blocked air filter

Oil fouling - Wet oily deposits leading to misfire and weak spark. Caused by oil leakage past piston rings or valve guides (4-stroke engine), or excess lubricant (2-stroke engine)

Overheating - A blistered white insulator and glazed electrodes. Caused by ignition system fault, incorrect fuel, or cooling system fault

Worn plug - Worn electrodes will cause poor starting in damp or cold weather and will also waste fuel

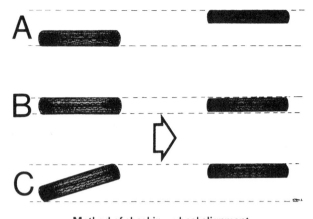

Method of checking wheel alignment

A and C - incorrect   B - correct

The electrode gap can be measured using feeler gauges. If necessary, alter the gap by bending the outer electrode, preferably with the proper electrode tool. *Never bend the centre electrode, otherwise the porcelain insulator will crack, and may cause damage to the engine if particles break away whilst the engine is running.* If the outer electrode is seriously eroded as shown in the photographs, or if the spark plug is heavily fouled, it should be renewed. Renew the spark plug annually regardless of its apparent condition, as it will have passed peak efficiency. Clean the electrodes using a wire brush or a sharp-pointed knife followed by rubbing a strip of fine emery across the electrodes. If a sand-blaster is used, check carefully that there are no particles of sand trapped inside the plug body to fall into the engine at a later date. For this reason such cleaning methods are no longer recommended; if a plug is that heavily fouled it should be renewed.

Before refitting a spark plug to the cylinder head, coat the threads sparingly with a graphited grease to aid future removal. Use the correct size spanner when tightening the plug, otherwise the spanner may slip and damage the ceramic insulator. The plug should be tightened by hand only at first and then secured with a quarter of a turn of the spanner so that it seats firmly on its sealing ring.

Never overtighten a spark plug otherwise there is a risk of stripping the threads from the cylinder head, especially as it is cast in light alloy. A stripped thread can be repaired without having to scrap the cylinder head by using a 'Helicoil' thread insert. This is a low-cost service, operated by a number of dealers.

2 Check the transmission oil level

If the machine has just been run wait for approximately 15 minutes for the oil level to settle before checking it. The oil level is checked by means of a level screw in the left-hand crankcase cover. On 125 and 150 models the screw is situated near the tip of the gear lever, and on 250, 251 and 300 models it is situated just in front of the left-hand footrest.

To check the oil level, first ensure the machine is on level ground and is standing absolutely upright on its wheels, then remove the level plug from the cover. If the oil level is correct, oil should slowly trickle out of the level hole. If not, remove the filler plug from the top of the right-hand

crankcase half and add oil of the specified type through the filler plug orifice until it starts to trickle out of the level plug. Note: if the type of oil in the transmission is not known, the oil must be drained and renewed as described under the two-yearly heading. Once the oil slows to a trickle refit the screw and tighten it securely, having first checked the condition of its sealing washer. Refit the filler plug to the top of the crankcase ensuring that it is pushed fully home.

3 Check and lubricate the chain

Since the chain is fully enclosed it will last longer and need less maintenance than one which is exposed to the elements. However, regular checks and lubrication should not be neglected. If the chain rollers look dry, the chain needs lubrication immediately. Do not allow it to run dry so that the links start to kink, or until traces of reddish-brown deposit can be seen on the side plates.

To lubricate the chain, it will be necessary to peel back one of the rubber gaiters, and retain it, or remove the right-hand crankcase cover; the latter method is probably the easiest. The chain should be lubricated by applying aerosol chain lubricant or gear oil evenly along its entire length until all the surfaces of the chain are covered. Then refit the gaiter or crankcase cover.

It is also necessary to check the chain tension at regular intervals to compensate for wear. Since this wear does not take place evenly along the length of the chain, tight spots will appear which must be compensated for when adjustment is made. Chain tension is checked with the machine on its centre stand, so that the rear wheel is raised clear of the ground, and the transmission in neutral. Find the tightest spot of the chain by rotating the rear wheel whilst pushing down on the upper chain gaiter, testing the entire length of the chain at least once. When the tightest spot has been found, the chain tension is checked at the mid-point of the upper chain gaiter. When the chain tension is correct, without using excessive force, it should be possible to press the gaiter down until its bottom edge just contacts the swinging arm. If not, slacken both the rear wheel spindle, sprocket retaining nut and the adjuster locknuts to permit adjustment. Slacken the adjuster nuts until the chain is too slack, then draw the wheel backwards by tightening the adjuster nuts until the tension is correct. To preserve accurate wheel alignment it is essential that both adjuster nuts are rotated by exactly the same amount. When adjustment is correct, tighten the rear wheel spindle, sprocket retaining nut and adjuster locknuts securely.

To check the accuracy of the wheel alignment lay a plank of wood (or draw a length of string) parallel to the machine so that it touches both walls of the rear tyre. Wheel alignment is correct when the plank or string is equidistant from both walls of the front tyre when tested on both sides of the machine, as shown in the accompanying illustration.

4 Check the fuel pipe and carburettor settings

Note: *petrol is extremely flammable, especially when in the form of vapour. Take all precautions to prevent the risk of fire and read the Safety first! section of this manual before starting work.*

Give the pipe which connects the fuel tap and carburettor a close visual examination and check for cracks or any signs of leakage. In time, the synthetic rubber pipe will tend to deteriorate and will eventually leak. Apart from the obvious fire risk, the leaking fuel will affect fuel economy. If the pipe is found to be damaged it must be renewed. Refer to Chapter 2 for further information.

Once the fuel pipe has been examined it is necessary to check the

Transmission oil level screw location - 125 and 150 models

On 250, 251 and 300 models transmission oil level screw is obscured by gearchange lever

If necessary, top up using the required oil

Adjust drive chain tension as described in text

Adjust oil pump cable until marks align at correct throttle position

throttle and choke cable freeplay. Although the manufacturer does not specify any freeplay measurements, as a guide it is recommended that there should be approximately 2 mm (0.08 in) of freeplay on each cable. If necessary, the cables can be adjusted using the adjusters situated on the carburettor, the throttle cable adjuster being situated on the threaded top, and the choke cable adjuster on the right-hand side of the carburettor body. To adjust the cables, slide the rubber cover up the cable and slacken the adjuster locknut. Rotate the adjuster until the correct amount of freeplay is obtained, then tighten its locknut securely and refit the rubber cover.

If rough running of the engine has developed, some adjustment of the carburettor pilot screw and idle speed may be required. If this is the case refer to Section 8 of Chapter 2. Do not make these adjustments unless they are obviously required; there is little to be gained from unwarranted attention to the carburettor.

Once the carburettor has been checked and adjusted as necessary, open and close the throttle several times, allowing it to snap shut under its own pressure. Ensure that it is able to shut off quickly and fully at all handlebar positions. If this is not the case, remove the screw which retains the right-hand handlebar end plug and remove the plug. Slacken the twistgrip housing clamping screw and dismantle the twistgrip assembly. Where fitted, remove the screw which retains the oil pump cable clip and disengage the oil pump cable from the twistgrip. Remove all traces of old grease and examine all the components for wear or damage, renewing as necessary. If required the throttle and oil pump cables (as applicable) can be lubricated as described under the annual

heading. Apply general purpose grease to all bearing surfaces of the twistgrip components and reassembly them by reversing the dismantling procedure, tightening all screws securely.

5 Check the oil pump pipes and cable adjustment – oil injection models

Remove the two screws which retain the oil pump cover and lift the cover away from the engine. Give both the oil pipes a close visual inspection for cracks and signs of leakage and ensure that they are secured at each end by their retaining clips. If a damaged pipe is discovered it must be renewed immediately. If the pipes are disturbed for any reason, do not forget to bleed the system, as described in Chapter 2, after reconnection.

Prior to checking the cable adjustment, check the throttle cable adjustment as described above. To check the oil pump cable adjustment, start the engine and warm it up thoroughly until it ticks over smoothly. Then slowly rotate the throttle twistgrip until the engine revs rise slightly and hold the twistgrip steady so that the engine is running at approximately 1200 – 1500 rpm. With the twistgrip held in this position the line on the oil pump pulley should align with the line on the pump body. If not, slide the rubber cover up the cable, slacken the adjuster locknut and rotate the adjuster until the two marks align. Open and close the throttle a few times, then recheck the setting. Repeat the adjustment procedure as described, then tighten the adjuster locknut securely and refit the rubber cover.

6 Overhaul the brakes

Note: *brake fluid will discolour or remove paint if contact is allowed. Avoid this where possible and remove accidental spillage immediately. Similarly, avoid contact with all plastic components.*

Front disc brake

To remove the brake pads, slacken both caliper mounting bolts and slide the caliper off the disc. Lever off the cover from the top of caliper and pull out the pad retaining pins. If stuck in place due to corrosion, the retaining pins can be tapped out of the caliper using a hammer and suitably sized drift. Once both pins have been removed, lift out the pad spring and withdraw both pads from the caliper. If the pads are fouled with grease or oil, or are heavily scored or damaged by dirt or debris, they must be renewed as a set; there is no satisfactory way of degreasing the friction material. Measure the thickness of the friction material of each pad. If either pad is worn close to or beyond the service limit given in the Specifications, both pads must be renewed as a set.

If the pads can be used again, clean them carefully using a fine wire brush that is completely free of oil or grease. Use a pointed instrument to dig out any embedded particles of foreign matter from the friction material. Any areas of glazing may be removed using emery cloth. Remove all traces of corrosion the pad retaining pins.

On reassembly, if new pads are to be fitted, the caliper pistons must be pushed fully back into their bores to provide the necessary clearance to accommodate the additional friction material. If any undue stiffness is encountered the caliper assembly should be dismantled for examination as described in Chapter 5. While pushing the pistons back, remove the reservoir cap, ventilation ring and diaphragm, and maintain a careful eye on the fluid level in the reservoir. If the reservoir has been overfilled, the surplus fluid will prevent the pistons returning fully and must be removed by soaking it up with a clean cloth whilst taking care to prevent fluid spillage.

Insert the pads into the caliper and refit the pad spring. Apply a smear of high melting-point grease to the pad retaining pins and insert them into the caliper, ensuring that they engage correctly with the brake pads and pad spring, and refit the caliper cover. Refit the caliper to the disc and tighten its mounting bolts securely.

Apply the brake lever gently and repeatedly to bring the pads firmly in contact with the disc until full brake pressure is restored. Be careful to watch the fluid level in the reservoir; if the pads have been re-used it will suffice to keep the level above the lower level mark, by topping up if necessary, but if new pads have been fitted the level should be slightly higher to allow for the pads to bed in. Refit the diaphragm and ventilation ring and tighten the reservoir cap securely.

Before taking the machine out on the road, be careful to check for fluid leaks from the system, and check that the front brake is functioning correctly. Remember that new pads, and to a lesser extent cleaned pads, will require a bedding-in period before they reach peak efficiency. Where new pads are fitted use the brake gently but firmly for the first 50 – 100 miles to enable the pads to bed fully in.

Drum brakes – front and rear

Since there is no external means of assessing brake wear, the wheels must be removed so that the brake components can be dismantled, cleaned, checked for wear and renewed as necessary. Refer to Chapter 5 for further information on overhauling drum brakes. On reassembly, apply a thin smear of high melting-point grease to the bearing surfaces of the brake camshafts and shoe pivots.

7 Check the wheels and wheel bearings

Place the machine on its centre stand so that the wheel to be examined is raised clear of the ground.

Examine the rim for serious corrosion or impact damage. Slight deformities can often be corrected by adjusting spoke tension, although serious damage or corrosion will necessitate renewal, which is best left to an expert.

Place a wire pointer close to the rim and rotate the wheel to check it for runout in the radial and axial planes. If the rim runs significantly out of true in either plane, check spoke tension by tapping them with a screwdriver. A loose spoke will sound quite different to those around it. Note that worn wheel bearings will also cause wheel runout.

Adjust spoke tension by turning the square-headed nipples with the appropriate spoke key which can be purchased from a dealer. With the spokes evenly tensioned, remaining distortion can be pulled out by tightening the spokes on one side of the wheel and slackening those on the opposite side. This will pull the rim across whilst maintaining spoke tension.

More than slight adjustment will cause the ends of the spokes to protrude through the nipple and chafe the inner tube, causing a puncture. Remove the tyre, inner tube and the rim band and file off any protruding ends of the spokes. The rim band protects the tube against chafing – check that it is in good condition before refitting it.

Spoke tension and general wheel condition must be checked regularly. Frequent cleaning will help prevent corrosion. If a broken spoke is found, it must be renewed immediately as the load taken by it will immediately be transferred to the adjacent spokes which in turn may fail.

An out of balance wheel will produce a hammering effect through the steering at high speed; check this as described in Chapter 5.

To check the wheel bearings, grasp each wheel firmly at the top and bottom and attempt to rock it from side to side; any freeplay indicates worn bearings which must be renewed as described in Chapter 5.

8 Check and lubricate the centre stand pivot

It is essential that the smooth, safe operation of the centre stand is ensured by regular lubrication, which will serve two purposes. The first and most important, is that a layer of grease will prevent the onset of corrosion which would cause the stand to become stiff and jerky in operation, and the second is that lubrication will minimise the wear which will inevitably occur on any bearing surface.

Dismantle, clean and grease the centre stand pivot as described in Section 14 of Chapter 4.

Six monthly, or every 3100 miles (5000 km)

Perform all the tasks listed under the previous headings, then carry out the following.

1 Clean the air filter

To remove the air filter, first unlock and remove the right-hand sidepanel. Disconnect the battery, ensuring that the negative (–) terminal is disconnected first, and remove it from the machine. Slacken and remove the nut which retains the air filter housing cover, and lift the cover and gasket away from the housing. Remove the nut and washer which retain the air filter element and withdraw both the retaining plate and element from the machine.

If the element is wet or damaged (check carefully for splits or tears in the paper pleats) or if it is so badly clogged with dirt that it cannot be cleaned effectively, it must be renewed.

If serviceable, tap the element lightly on a hard surface to dislodge any particles of dirt, then clean by blowing from the inside outwards with compressed air. Wipe the inside of the filter housing and cover

Position brake pads ...

... and pad spring in caliper ...

... and insert pad retaining pins

Refit caliper cover ...

... and refit caliper to disc, tightening its mounting bolts securely

Refit the element ...

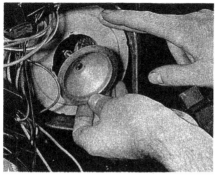

... and retaining plate to the air filter housing ...

... and secure them in position with their nut and washer

Refit the air filter housing cover and gasket and tighten its retaining nut securely

Adjusting contact breaker points

Slacken retaining screws and move contact breaker baseplate to adjust ignition timing

clean and examine the gasket for signs of damage, renewing it if necessary.

On reassembly apply a film of grease to the element edges to help achieve a good seal, refit the element retaining plate and washer and tighten its retaining nut securely. Refit the gasket and housing cover, ensuring that they are correctly seated, and tighten the cover retaining nut securely. Refit the battery, remembering to connect the negative (–) terminal last. Refit the sidepanel.

Note that the filter element and cover must be positioned correctly to provide a good seal so that unfiltered air cannot carry dirt into the engine. Additionally, note that the carburettor is jetted to compensate for the presence of the air filter; serious engine damage will occur through overheating as a result of a weak mixture if the element is damaged or bypassed.

This interval is the maximum for filter cleaning; if the machine is used in wet weather or in very dirty or dusty conditions, the filter must be cleaned more frequently.

2 Check the contact breaker points and ignition timing

On 125 and 150 models remove the circular cover from the right-hand crankcase cover, and on 250, 251 and 300 models remove the right-hand crankcase cover. Remove the spark plug. The contact breaker gap and ignition timing are closely inter-related and their adjustment should therefore be considered as one task. As many owners may not possess a DTI (Dial test indicator), required to check the timing with the necessary degree of accuracy, they may prefer to have this check carried out by an authorized MZ dealer. For ease of reference, the various operations of this check are laid out below under separate sub-headings.

Checking the contact breaker points

Using a spanner on the generator rotor bolt, turn the engine over until the contact breaker points are fully open. Check the points for pitting and burning. If this is only slight it can be removed with a small oil stone or emery paper, although care should be taken to keep the contact faces square. If wear is severe renew the contact breaker assembly.

To remove the contact breaker, undo the nut and lift off the connecting lead to the condenser. Remove the single screw which retains the contact breaker assembly and remove it from the machine.

On refitting, lightly oil the contact breaker pivot pin and tighten the condenser lead retaining nut securely.

Soak the lubricating pad in a hypoid oil and squeeze out the excess before refitting. The felt should be adjusted so that it is just in contact with the lobe of the cam. Do not oil the felt wiper pad attached to the contact breaker; this is fitted to keep surplus oil off the points.

Adjusting the contact breaker points

Turn the crankshaft using a spanner on the generator rotor bolt until the raised lobe of the cam is bearing against the contact breaker heel, and the points are open to their fullest extent. Check this very carefully and repeat if necessary as it is essential that the gap is set with the points fully open.

If the gap is correct a feeler gauge of 0.3 mm (0.012 in) thickness should be a light sliding fit between the points. If adjustment is necessary, slacken the single retaining screw slightly and rotate the eccentric adjusting screw until the gap is correct. Then tighten the retaining screw securely and recheck the setting.

The contact breaker gap must be accurately set if optimum performance is to be achieved. If the gap is too small it will retard the ignition setting and if too large, the ignition will be over-advanced.

Checking the ignition timing

Before the ignition timing can be checked, the contact breaker points must be cleaned, and the gap correctly set as described above.

The timing can only be checked statically, and since timing marks are not provided some means must be found of establishing the exact point at which the spark must take place. This is best achieved using a dial gauge and adaptor to establish the piston position. In addition to this, a method must be devised to determine accurately exactly when the points open. The simplest method of doing this is to connect a test lamp between the condenser terminal and a good earth point; with the ignition switched on the lamp will light as the points open.

Screw the dial gauge adaptor into the spark plug thread, fit the ball-tipped extension to the gauge and insert it into the adaptor, tightening the retaining grub screw. Turn the crankshaft forwards (clockwise); the gauge reading will decrease, stop momentarily as TDC

(Top Dead Centre) is reached, and then increase again. Find the TDC position, zero the gauge and then turn the crankshaft backwards until a reading of 4 - 5 mm is obtained. Switch on the ignition and turn the crankshaft forwards until the gauge shows a reading of 2.5 mm (0.10 in) before TDC; at this point the lamp should light, indicating that the points have opened.

If adjustment is necessary, slacken the two screws which retain the contact breaker baseplate and very carefully move the plate until the lamp lights at the correct point; moving the plate to the left advances the timing and moving it to the right retards the timing. Tighten the baseplate retaining screws and recheck the setting. When the timing is correct, remove the test equipment and refit the spark plug and engine cover.

3 Clean the fuel tap filters

Note: *Petrol is extremely flammable, especially when in the form of vapour. Take all precautions to prevent the risk of fire and read the Safety first! section of this manual before starting work.*

The manufacturer recommends that the fuel tank is drained and the tap removed, dismantled and cleaned at this interval. While this is a task which must be carried out at regular intervals, it should be unnecessary to do so with such frequency if care is taken to prevent dirt or water from entering the fuel tank. It should prove sufficient to unscrew the tap filter bowl and to examine its contents. If there is no more than a trace of dirt or water in the bowl and the lower filter is clean, there is no need to dismantle the tap. However, if there is a great deal of foreign matter in the filter bowl and the filter is dirty or clogged, the tap should be dismantled and cleaned as described in Section 3 of Chapter 2, and the fuel tank flushed out.

4 Clean the carburettor

Note: *Petrol is extremely flammable, especially when in the form of vapour. Take all precautions to prevent the risk of fire and read the Safety first! section of this manual before starting work.*

The manufacturer recommends that the carburettor should be removed and cleaned at this interval. Provided the machine starts and runs well with no signs of carburation faults, it is suggested that this too is unnecessary at quite so frequent intervals, there is little point in disturbing the carburettor unnecessarily unless dirt was found in the fuel during the previous operation. The carburettor must then, of course, be dismantled and cleaned as described in Chapter 2.

Under normal circumstances carburettor maintenance can be restricted to checking the mountings and carburettor top for security and examining the air filter hose for signs of damage such as cracks and splits. Also ensure that the hose is securely fitted to the carburettor by its clamp.

5 Check the electrical connections

At regular intervals all electrical connections should be checked for cleanliness and security. Work methodically around the machine, checking spade terminals for security and screw terminals for tightness, and that all electrical components are clean and securely fastened. Check the wiring harness for any signs of damage and ensure that all bulb and switch contacts are clean. If damage is found at any point, it must be repaired or the relevant component renewed. All traces of dirt and corrosion must be polished away from the terminals to minimise power loss. Apply WD-40 or a similar water-dispersant fluid to all exposed switches to prevent corrosion of the terminals.

6 General check

Work methodically around the machine, checking all nuts and bolts for tightness, paying particular attention to the engine mountings, rear suspension and front fork mounting bolts. Ensure all fasteners are tightened securely but be careful not to over tighten any component. Where possible make use of the torque wrench settings given in this Manual.

Annually, or every 6200 miles (10 000 km)

Carry out all the tasks listed under the previous headings, then carry out the following:

1 Renew the air filter

Following the instructions given under the six monthly heading, remove the air filter element, discard it and fit a new one. This must be done irrespective of the element's condition to preserve the machine's economy and performance.

2 Renew the spark plug

To preserve the machine's economy, reliability and performance the spark plug must be renewed at this interval, regardless of its apparent condition, as it will have passed peak efficiency. Check that the new plug is of the correct type and heat range and that it is gapped correctly before it is fitted. Coat the threads of the plug with a graphited grease prior to installation to aid future removal, and tighten the plug first by hand and then by another quarter of a turn using a suitable plug spanner.

3 Check and grease the chain

While the regular checks described previously will maintain the chain in good condition, it must be removed at this interval to be cleaned, checked for wear and thoroughly greased.

If the chain alone is to be removed, the simplest method is to remove the right-hand crankcase cover, disconnect the chain at its connecting link and connect a worn-out chain of equal size and length to one end. By pulling on the other end the worn out chain will pass through the gaiters and around the rear sprocket until the original chain can be disconnected from it and removed. On refitting the process is reversed to install the freshly lubricated chain. However, since the rear wheel and cush drive have to be dismantled at this interval to carry out other maintenance tasks, it is preferable to remove the sprocket cover and chain gaiters as well. This does not take much longer and gives an opportunity for a complete examination of all components.

Immerse the chain in a bath containing a mixture of petrol and paraffin and use a stiff-bristled brush to scrub away all traces of dirt and old grease. *Note that petrol is extremely flammable, especially when in the form of vapour. Take all precautions to prevent the risk of fire.* Swill the chain around to ensure the solvent penetrates fully into the bushes and rollers and removes any lubricant which may still be present. When the chain is completely clean, remove it from the bath and hang it up to dry.

To check the chain, it must be cleaned and dried as described above, then laid out on a flat surface. Compress the chain fully and measure its length from end to end. Anchor one end of the chain and pull on the other end, drawing the chain out to its fullest extent. Measure the stretched length. If the stretched length exceeds the compressed measurement by more than 2%, or $\frac{1}{4}$ in per foot, the chain is worn out and should be renewed. Additionally, if any links are kinked or stiff through lack of lubrication, or if such damage as split or missing rollers or side plates is found, the chain must be renewed.

Chain renewal should always be done in conjunction with both sprockets since the running together of new and worn components greatly increases the rate of wear. Similarly, check both sprockets for wear. If either is found to be damaged, or if its teeth are hooked, chipped or worn, it must be renewed. It is false economy to renew only one sprocket at a time, for the reason given above, therefore both sprockets and the chain should be renewed as a set.

If the chain is still serviceable it must be lubricated thoroughly so that the grease reaches all the inner bearing surfaces. This can be done only when the chain has been removed, cleaned and dried as described above, then by immersing it in a molten bath of special chain lubricant such as Chainguard or Linklyfe. Follow carefully the manufacturers' instructions when using Chainguard or Linklyfe and take great care to swill the chain gently in the molten lubricant to ensure that it penetrates all bearing surfaces. Wipe off the surplus lubricant before refitting the chain to the machine.

Before refitting the chain, check that the rubber gaiters are not damaged or worn. These not only exclude dirt and water but also serve to guide the chain and prevent it from whipping. Renew the gaiters if either is split or torn, or if their internal ribs are severely worn.

On refitting, ensure that the chain's connecting link spring clip is fitted with its closed end facing the normal direction of chain travel.

4 Check the suspension and steering

Place the machine on its centre stand and place blocks beneath the crankcase so that the front wheel is raised clear of the ground.

Grasp the front fork legs near the wheel spindle and push and pull firmly in a fore and aft direction. If play is evident between the top and bottom yokes and the frame, the steering head bearings are damaged or worn and must be renewed as described in Chapter 4. *Do not confuse steering head bearing play with any play which may be evident in the front forks.* Finish the steering check by placing the handlebars in the straightahead position and tapping lightly on one end; the forks should fall smoothly from lock to lock, with no signs of stiffness in the bearings, although some may be evident due to the drag of the cables and wiring.

At the same time as the steering head bearings are checked, take the opportunity to examine closely the front suspension. Ensure that the front forks work progressively and smoothly by pumping them up and down whilst the front brake is held on. Any faults revealed by this check should be investigated further, as any deterioration in stability of the machine can have serious consequences. Check carefully for signs of leakage around the front fork oil seals. If any damage is found, it must be repaired immediately as described in the relevant Sections of Chapter 4. If all is well check the front fork oil level as follows.

To check the front fork oil the machine must be placed on its centre stand and the front wheel raised clear of the ground so that the forks are fully extended, and at least 2 hours must be allowed from the time the machine was last ridden to allow the oil level to settle. Remove the top plugs from the fork legs and obtain a length of 4 mm (0.16 in) diameter welding rod or something similar which is at least 6 inches longer than the fork legs. Insert this into the fork leg and slide it down until it touches the bottom of the fork leg (a distance which can be checked by holding the rod against the outside of the fork).

The depth measured on the rod should be as specified and must be exactly the same in both fork legs; if not add oil of the required type until the level is correct. Refit the fork top plugs, having first applied a smear of jointing compound to their threads, and tighten them to the specified torque setting.

Push the machine off the centre stand and check that the rear suspension functions smoothly and progressively. Then place the machine on its centre stand and check for wear in the swinging arm pivot by pushing and pulling horizontally at its rear end, noting that there should be no discernible free play. At regular intervals the swinging arm pivot should be removed, cleaned and greased. This essential to prevent the formation of corrosion which would otherwise render the shaft immovable. Refer to Chapter 4 for further information on all operations involving the rear suspension.

5 Grease the wheel bearings and speedometer drive

Referring to Chapter 5, take out each wheel in turn so that the above components can be removed, cleaned, checked for wear and renewed if necessary. On reassembly, pack the wheel bearings and speedometer drive with fresh high melting-point grease.

6 Lubricate the control and instrument drive cables

Check the outer cables for signs of damage, then examine the exposed portions of the inner cables. Any sign of kinking or fraying will indicate that renewal is necessary. To obtain maximum life and reliability from the cables they should be thoroughly lubricated. To do the job properly and quickly use one of the hydraulic cable oilers available from most motorcycle shops. Free the upper end of the cable and assemble the oiler as described by the manufacturer's instructions. Operate the oiler until oil emerges from the lower end of the cable, indicating that the cable is lubricated throughout its length. This process will expel any dirt or moisture and will prevent its subsequent ingress.

If a cable oiler is not available, an alternative is to remove the cable from the machine. Hang the cable upright and make a small funnel arrangement using plasticine or by tapping a plastic bag around the upper end of the cable. (See accompanying illustration.) Fill the funnel with oil and leave it overnight to drain through. Note that where nylon-lined cables are fitted, they should be used dry or lubricated with a silicone-based lubricant suitable for this application. On no account use ordinary engine oil because this will cause the liner to swell, pinching the cable.

To ensure the correct and accurate operation of the speedometer and tachometer (where fitted), the drive cables must be removed from the machine and examined as described in Section 15 of Chapter 4. Clean the inner cable with a rag moistened in solvent, then regrease the cable ensuring that no grease is applied to the upper six inches of the

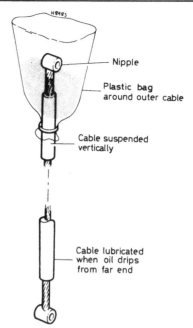

Nipple

Plastic bag around outer cable

Cable suspended vertically

Cable lubricated when oil drips from far end

Oiling a control cable

Refit spring clip with closed end in normal direction of travel

Checking the fork oil level

Transmission oil drain plugs – 125 and 150 models

On 250, 251 and 300 models drain plugs are situated on underside of the crankcase ...

... and at bottom of clutch cover

inner cable. If this precaution is not observed, grease will work its way into the instrument and immobilise the sensitive movement.

When refitting a drive cable, always ensure that it has a smooth run to minimise wear, and check that it is secured by any necessary clamps or ties.

Two yearly, or every 12 400 miles (20 000 km)

Carry out all the tasks listed under the previous headings, then carry out the following:

1 Change the transmission oil

Oil changes are much quicker and more efficient if the machine is taken for a journey long enough to warm the oil up to normal operating temperature so that it is thin and is holding any impurities in suspension.

Place a suitably-sized container below the engine and remove both drain plugs. On 125 and 150 models the drain plugs are the two forwardmost bolts on the underside of the crankcase. On 250, 251 and 300 models they comprise of the right-hand of the two bolts on the underside of the crankcase, and the screw at the bottom of the left-hand crankcase cover. On all models remove the filler plug from the top of the crankcase. Whilst waiting for the oil to drain, examine the drain plug sealing washers and renew them if damaged. When the oil has drained completely, refit both drain plugs, tightening them securely, and wipe away any surplus oil.

The nearest equivalent available in the UK to the gearbox oil specified by the manufacturer is a hypoid EP80 gear oil. Although it is stated that it is permissible to use an ordinary engine oil of SAE30 or 40 viscosity. The UK importers have found EP80 to be satisfactory in normal use.

Refill the gearbox with the specified amount of oil, then check the level as described under the three-monthly heading. When the oil level is correct, refit the level and filler plugs and wipe off any spilt oil.

2 Renew the brake fluid (disc brake models only)

It is necessary to renew the brake fluid at this interval to preserve maximum brake efficiency by ensuring that the fluid has not been contaminated and deteriorated to an unsafe level.

Before starting work, obtain a new, full can of the recommended hydraulic fluid and read carefully the Section on brake bleeding in Chapter 5. Prepare the clear plastic tube and glass jar in the same way as for bleeding the brake, open the bleed nipple by unscrewing it $\frac{1}{4} - \frac{1}{2}$ a turn with a spanner and apply the front brake lever repeatedly; this will pump out the old fluid. *Keep the master cylinder topped up at all times, otherwise air may enter the system and greatly lengthen the operation.* The old brake fluid is invariably much darker in colour than the new, making it easier to see when it is pumped out and the new fluid has completely replaced it.

When the new fluid appears in the clear plastic tubing completely uncontaminated by traces of old fluid, close the bleed nipple, remove the plastic tubing and refit the rubber cap on the nipple. Top up the master cylinder reservoir to above the lower level mark, unless the brake pads have been renewed in which case slightly more fluid should be added. Clean and dry the rubber diaphragm then refit it along with the ventilation ring and tighten the reservoir cap securely.

Wash off any surplus fluid and check that the brake is operating correctly before taking the machine out on the road.

3 Change the front fork oil

It is essential that the fork oil is changed at this interval as it deteriorates in use and can become contaminated by dirt, water and metallic particles. Over a long period this can lead to a reduction in fork damping which could seriously affect the machine's handling.

Since drain plugs are not fitted to the lower fork legs, the forks must be removed from machine and inverted to pump out the old oil. Refer to Chapter 4 for details of fork removal and draining. Refit the forks, filling them with the correct amount of the specified oil, and then check the fork oil level as described under the annual heading.

Additional routine maintenance

Certain tasks do not fall under the previous mileage/time headings as they concern items which deteriorate with age, whether the machine is used a great deal or hardly at all.

1 Decarbonisation

The oily nature of any two-stroke engine's exhaust leads to layers of carbon being deposited in the combustion chamber and exhaust system. If not removed at regular intervals these deposits will build up to the point where the machine's performance and economy are significantly reduced.

It is very difficult to give a precise interval for decarbonisation as so many different factors have to be taken into account. For example if maintenance is neglected so that deposits build up at a faster rate through inefficient combustion, or if the wrong type of engine oil is used. Furthermore, the rider's driving style must be taken into account; any machine that is used principally for fast riding or for long journeys on open roads will not require decarbonisation as often as a machine which is used for short, low-speed commuting trips. Some machines will therefore require decarbonisation once a year while others will run satisfactorily for more than twice as long before attention is required. As an initial starting point it is suggested that this task be carried out once a year until sufficient experience has been gained to either lengthen or shorten the interval as necessary. Note that it may not be necessary to decarbonise the exhaust as frequently as the engine, or vice versa, so the two can be treated as separate items with their own schedules.

Cylinder head and barrel

Refer to Chapter 1 and remove the cylinder head and barrel. It is necessary to remove all carbon from the head, barrel and piston crown whilst avoiding removal of the metal surface on which it is deposited. Take care when dealing with the soft alloy head. Never use a steel scraper or a screwdriver. A hardwood, brass or aluminium scraper is ideal as these are harder than the carbon but not harder than the underlying metal. With the bulk of the carbon removed, use a brass wire brush. Finish the head and piston with metal polish; a polished surface will slow the subsequent build up carbon. Clean out the barrel ports to prevent the restriction of gas flow. Remove all debris by washing each component in paraffin whilst observing the necessary fire precautions. Renew the piston rings, if necessary, on reassembly.

Exhaust system

First check that the exhaust system is securely fastened and that there are no exhaust leaks. Some idea of the condition of the exhaust can be gained by looking at the end of the silencer. If the petroil mixture or oil pump settings (as applicable) are known to be correct, and the machine is ridden normally, there should be a thin film of sooty black carbon with a slight trace of oiliness. If the machine is ridden hard, the deposits will tend to be a lighter colour, almost grey, and there will be no trace of oil. If any of the above is found, and the spark plug electrode colour, carburettor settings, air filter condition and, if necessary, the oil pump settings are known to be satisfactory from the previous Routine maintenance operations, then the carbon deposits inside the exhaust system will be kept to a minimum and will take a long time to build up to the point where a full decarbonisation operation is necessary. On the other hand, if the rear of the exhaust is excessively oily and the carbon deposits rather thicker than those described above, then either one of the settings mentioned above is incorrect, causing the engine to run inefficiently so that it produces too much waste in the form of carbon, or the machine is being ridden too slowly which produces the same symptoms. If the engine settings are known to be satisfactory thanks to the previous Routine maintenance operations and the carbon build-up is caught at an early stage, then the simplest and most satisfying method of cleaning the exhaust is to take the machine on a good hard run until the exhaust is too hot to touch and the excessive smoke produced by the dispersal of the carbon/oil build-up has disappeared. This is a well-known trick employed by many mechanics to restore lost power, and can have dramatic results. If the build-up has been allowed to develop too much, however, the complete exhaust must be removed from the machine and decarbonised using one of the methods described below.

Two possible methods of cleaning exist; the first, which will only be really effective if the deposits are very oily, consists of flushing the system with a petrol/paraffin mixture. It will be evident that great care must be taken to prevent the risk of fire when using this method, and the system must be hung overnight in such a way that all the mixture will drain out of the exhaust. It will also be necessary to use a suitable scraping tool to remove any hardened deposits from the front length of the exhaust pipe and from the exhaust port of the barrel.

The second method of exhaust cleaning is to use a solution of caustic soda to dissolve the deposits. While this is a lengthy and time-consuming operation, it is the simplest and most effective method that can be used by the average owner, and is therefore described in detail in the following paragraphs. Remove the exhaust system from the machine and suspend it from its silencer end. Block up the end of the exhaust pipe with a cork or wooden bung. If wood is used, allow an outside projection of three or four inches with which to grasp the bung for removal. Bear in mind that it is very important to take great care when using caustic soda as it is a very dangerous chemical. Always wear protective clothing, this must include proper eye protection. If the solution does come into contact with the eyes or skin it must be washed off immediately with clean, fresh, running water. In the case of an eye becoming contaminated, seek expert medical advice immediately. Also, the solution must not be allowed to come in contact with

aluminium alloy – especially at the above recommended strength – caustic soda reacts violently with aluminium and will cause severe damage to the component.

The mixture to be used is a ratio of 3 lbs caustic soda to a gallon of fresh water. This is the strongest solution ever likely to be required. Obviously the weaker the mixture the longer the time required for the carbon to be dissolved. Note, whilst mixing the solution, that the caustic soda should be added to the water gradually, whilst stirring. Never pour water into a container of caustic soda powder of crystals; this will cause a violent reaction to take place which will result in great danger to one's person.

Commence the cleaning operation by pouring the solution into the system until it is quite full. Do not plug the open end of the system. The solution should not be left overnight for its dissolving action to take place. Note the solution will continue to give off noxious fumes throughout its dissolving process; the system must therefore be placed in a well ventilated area. After the required time has past, carefully pour out the solution and flush the system through with clean, fresh water. The cleaning operation is now complete.

2 Renew the brake caliper and master cylinder seals – disc brake models only

In the interest of safety, it is recommended that the brake caliper and master cylinder seals should be renewed at two-yearly intervals. The brake caliper and master cylinder can be overhauled as described in Chapter 5.

3 Renew the hydraulic brake hose – disc brake models only

In the interest of safety, it is also recommended that the brake hose should be renewed every four years, regardless of its apparent condition. Refer to Chapter 5 for details.

4 Cleaning the machine

Keeping the motorcycle clean should be considered as an important part of routine maintenance, to be carried out whenever the need arises. A machine cleaned regularly will not only succumb less speedily to the inevitable corrosion of external surfaces, and hence maintain its market value, but will be far more approachable when the time comes for maintenance or service work. Furthermore, loose or failing components are more readily spotted when not partially obscured by a mantle of road grime and oil.

Surface dirt should be removed using a sponge and warm, soapy water; the latter being applied copiously to remove the particles of grit which might otherwise cause damage to the paintwork and polished surfaces.

Use a wax polish on the painted parts. The plated parts of the machine should require only a wipe with a damp rag. If they are badly corroded, as may occur during the winter months, when the roads are salted, it is preferable to use one of the proprietary chrome cleaners. These often have an oily base, which will help to prevent the corrosion from recurring.

If the engine parts are particularly oily, use a cleaning compound such as 'Gunk' or 'Jizer'. Apply the compound whilst the parts are dry and work it in with a brush so that it has the opportunity to penetrate the film of grease and oil. Finish off by washing down liberally with plenty of water, taking care that it does not enter the carburettor, air filter or the electrics.

Whenever possible, the machine should be wiped down after it has been used in the wet, so that it is not garaged under damp conditions which will promote rusting. Remember there is little chance of water entering the control cables and causing stiffness of operation if they are lubricated regularly as recommended in the preceding sections.

Chapter 1 Engine, clutch and gearbox

Refer to Chapter 7 for information relating to 1991-on models

Contents

Specifications

Engine

Capacity:	
125 models	123 cc (7.6 cu in)
150 models	143 cc (8.8 cu in)
250 and 251 models	243 cc (15.0 cu in)
300 models	296 cc (18.1 cu in)
Bore:	
125 models	52.0 mm (2.05 in)
150 models	56.0 mm (2.20 in)
250 and 251 models	69.0 mm (2.72 in)
300 models	75.5 mm (2.97 in)
Stroke:	
125 and 150 models	58.0 mm (2.28 in)
250, 251 and 300 models	65.0 mm (2.56 in)
Compression ratio:	
125 and 150 models	10:1
250, 251 and 300 models	10.5:1

Cylinder barrel

Piston to bore clearance:	
125 and 150 models	0.03 mm (0.0011 in)
250 and 251 models	0.04 mm (0.0015 in)
300 models	0.035 mm (0.0013 in)

Piston

Note, specifications for 300 model were not available at the time of writing
Standard OD:
  125 models .... 51.96 – 51.99 mm (2.0456 – 2.0468 in)
  150 models .... 55.96 –55.99 mm (2.2031 – 2.2043 in)
  250 and 251 models .... 68.94 – 68.97 mm (2.7142 – 2.7153 in)
  300 models .... Not available
Ring groove width:
  Top ring:
    125 and 150 models .... 2.06 – 2.08 mm (0.0811 – 0.0818 in)
    250 and 251 models .... 2.08 – 2.10 mm (0.0818 – 0.0826 in)
  Central* and lower ring .... 2.04 – 2.06 mm (0.0803 – 0.0811 in)
  Service limit – all grooves... 2.10 mm (0.0826 in)
  250 and 251 models only

Piston rings

Note, specifications for 300 model were not available at the time of writing.
Ring height – all rings... 1.990 – 2.022 mm (0.0783 – 0.0796 in)
Service limit... 1.90 mm (0.0748 in)
End gap – installed .... 0.2 mm (0.008 in)
Service limit... 1.6 mm (0.063 in)

Crankshaft

Runout limit:
  125 and 150 models .... 0.02 mm (0.0008 in)
  250, 251 and 300 models .... 0.03 mm (0.0012 in)
Big-end axial clearance:
  125 and 150 models .... 0.210 – 0.523 mm (0.008 – 0.0206 in)
  250, 251 and 300 models .... 0.170 – 0.563 mm (0.007 – 0.022 in)
  Service limit – all models... 1.0 mm (0.039 in)
Big-end radial clearance .... 0.02 – 0.035 mm (0.0008 – 0.0013 in)
Service limit... 0.05 mm (0.0020 in)

Primary drive

Reduction ratio:
  125 and 150 models 2.055:1 (37/18T)
  250, 251 and 300 models 2.430:1 (68/28T)
Standard gear backlash – 250, 251 and 300 models only.................... 0.036 – 0.131 mm (0.0014 – 0.0052 in)
Service limit... 0.25 mm (0.0098 in)

Clutch

Number of friction plates:
  125 and 150 models 6
  250, 251 and 300 models 5
Number of plain plates:
  125 and 150 models 5
  250, 251 and 300 models 4
Number of clutch springs:
  125 and 150 models 1
  250, 251 and 300 models 6
Friction plate thickness:
  125 and 150 models .... 3.3 – 3.5 mm (0.130 – 0.138 in)
  Service limit... 3.2 mm (0.126 in)
  250, 251 and 300 models .... 2.9 – 3.1 mm (0.114 – 0.122 in)
  Service limit... 2.7 mm (0.106 in)
Plain plate thickness – all models .... 1.4 – 1.5 mm (0.055 – 0.059 in)
Distortion limit... 0.2 mm (0.008 in)
Clutch spring free length – 250, 251 and 300 models only .... 27.7 – 28.9 mm (1.09 – 1.14 in)

Gearbox

	125 and 150 models	250, 251 and 300 models
Ratios:		
1st	3.833* (34/12T)	3.000 (36/12T)
2nd	2.345* (26/15T)	1.865 (28/15T)
3rd	1.567* (22/19T)	1.333 (24/18T)
4th	1.191* (22/25T)	1.048 (22/21T)
5th	direct through shaft	0.870 (20/23T)

Note, on 125 and 150 models gear ratios from 1st – 4th are calculated taking into account that the layshaft output pinion (17T) and mainshaft 5th gear pinion (23T) are also used to transmit the drive to the sprocket.

Final drive

Reduction ratios:
  125 and 150 models 3.20:1 (48/15T)
  250 models 2.52:1 (48/19T)
  251 models 2.29:1 (48/21T)
  300 models 2.40:1 (48/20T)

Chain size... 428 ($\frac{1}{2}$ x $\frac{5}{16}$in)
Number of links:
    125, 150 and 251 models 128
    250 and 300 models .... 130 (128 with sidecar)

Torque settings

Component	kgf m	lbf ft
Cylinder head nuts:		
125 and 150 models	2.5	18.0
250, 251 and 300 models	2.6	19.0
Crankcase and crankcase cover screws:		
125 and 150 models	1.0	7.2
250, 251 and 300 models	1.3	9.5
Crankshaft oil seal cap screws – 125 and 150 models only	0.6	4.3
Main/output shaft oil seal cap screws – all models................................	0.5	3.6
Clutch centre nut – 125 and 150 models only	6.0 – 7.5	43.2 – 54.0
Clutch retaining nut/tachometer drive gear – 250, 251 and 300 models only..	8.0 – 10.0	57.6 – 72.0
Clutch spring retaining bolts – 125 and 150 models only.....................	0.5	3.6
Primary drive gear nut – 250, 251 and 300 models only.......................	6.0	43.2
Primary drive sprocket bolt – 125 and 150 models only	4.5 – 5.6	32.4 – 40.3
Generator rotor bolt	1.8 – 2.2	13.0 – 16.0
Generator stator screws	0.35 – 0.45	2.5 – 3.2
Engine mounting nuts and bolts:		
125 and 150 models	Not available	
250, 251 and 300 models	2.6	19.0

1   General description

All models employ an air-cooled, single cylinder, two-stroke engine, built in unit with the primary drive, clutch and gearbox. The cylinder head and barrel are of light alloy construction, the cylinder barrel being fitted with a cast iron liner. The crankshaft is of conventional design, having needle roller bearings at the connecting rod big- and small-ends and is supported by two ball journal main bearings.

On 125 and 150 models primary drive is via a chain, which is driven from a sprocket mounted on the left-hand end of the crankshaft, to the wet multi-plate clutch mounted on the end of the gearbox mainshaft. The five-speed gearbox is of the cross-over type, with the clutch and final drive sprocket being mounted on opposite ends of the mainshaft.

On 250, 251 and 300 models the wet multi-plate clutch unit is mounted on the left-hand of the crankshaft. Primary drive is transmitted through a pair of helical-cut gears; a small drive gear mounted on the crankshaft, and driven by the clutch assembly, transmits drive to a large driven gear mounted on the end of the input shaft. The five-speed gearbox is of conventional design, with the primary driven gear mounted on the input shaft and the final drive sprocket on the output shaft.

On all models the gearbox is lubricated by oil bath shared with the primary drive and clutch, the oil being contained in a reservoir formed by the main crankcase castings.

2   Operations with the engine/gearbox unit in the frame

The following components can be dismantled with the engine/gearbox unit in the frame:

  (a) Cylinder head, barrel and piston
  (b) Generator and ignition components
  (c) Clutch assembly and operating mechanism
  (d) Kickstart mechanism
  (e) Primary drive and tachometer drive (where fitted)
  (f) Crankshaft and mainshaft oil seals – 125 and 150 models
  (g) Output shaft oil seal only – 250, 251 and 300 models

3   Operations requiring engine/gearbox removal from the frame

It is necessary to remove the engine/gearbox unit from the frame

and separate the crankcase halves to gain access to the following components:

  (a) Crankshaft and main bearings
  (b) Crankshaft oil seals – 250, 251 and 300 models (See Section 14)
  (c) Gearbox and gear selector components

4   Removing the engine/gearbox unit from the frame

1   If the machine is dirty, wash it thoroughly before starting any major dismantling work. This will make work much easier and will rule out the risk of caked-on lumps of dirt falling into some vital component.

2   Drain the transmission oil as described in Routine maintenance. While the oil is draining remove the right-hand sidepanel and disconnect the battery (negative terminal first) to prevent the risk of any short circuits. If the machine is to be out of service for some time, remove the battery and give it regular refresher charges as described in Chapter 6.

3   Note that whenever any component is removed, all mounting nuts, bolts or screws should be refitted in their original locations with their respective washers and mounting rubbers and/or spacers.

4   Work is made much easier if the machine is raised to a convenient height on a purpose-built ramp or platform constructed of planks and concrete blocks. Ensure that the wheels are chocked with wooden blocks so that the machine cannot move and that it is securely tied down so that it cannot fall.

5   Remove the exhaust system and carburettor as described in Chapter 2.

6   Pull off the spark plug cap. Slacken and remove all the right-hand crankcase cover retaining screws and remove the cover from the machine. Disconnect the wires from the generator terminals noting exactly how they are arranged, and from the neutral switch (where fitted), and remove the wiring loom from the crankcase. Tie the wiring to the frame so that it is clear of the engine and will not hinder engine removal.

7   Straighten the gearbox sprocket locking tab and slacken the sprocket retaining nut whilst applying the rear brake to prevent the sprocket rotating (engine in gear). Slide the sprocket off the output shaft, noting that it may be necessary to slacken the drive chain to permit removal, and disengage the sprocket from the chain. On 125 and 150 models withdraw the spacer from the mainshaft oil seal and the clutch pushrod from the crankcase and store them with the casing for safekeeping.

8   Unscrew the knurled ring which secures the tachometer cable (where fitted) to its drive and disconnect it from the engine. On 250, 251 and 300 models, slacken the three screws which retain the tachometer

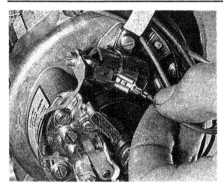

4.6a Disconnect wiring from stator housing terminals ...

4.6b ... and neutral switch (where fitted)

4.7 On 125 and 150 models withdraw spacer from mainshaft oil seal

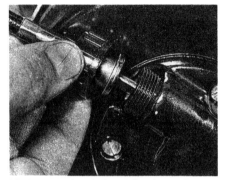

4.8a Unscrew knurled retaining ring and disconnect tachometer cable – 251 shown

4.8b On 250, 251 and 300 models remove O-ring and clutch adjusting plate

4.11 On 125 and 150 models remove rubbber plugs to reveal rear engine mounting bolts

drive housing or cover (as applicable) to the left-hand crankcase cover and remove it from the engine unit. Displace the rubber dust cover from the clutch cable mounting sleeve, situated on the front of the left-hand crankcase cover, then withdraw the clutch cable and remove its retaining spacer. Unscrew the mounting sleeve from the crankcase cover and disconnect the cable. Slide off the mounting sleeve and allow the cable to hang down from the front of the frame. Remove the clutch adjusting plate and O-ring from the left-hand crankcase cover and store them with the other components for safekeeping.

9   On all models, where fitted, remove the oil pump as described in Section 12 of Chapter 2.

10   The engine/gearbox unit should now be retained only by its mounting bolts. Make a final check that all components which might hinder the removal of the unit have been withdrawn.

11   Slacken and remove the two nuts and spring washers which secure the cylinder head to its upper mounting point, and slacken both the upper and lower rear engine mounting bolts. Note that on 125 and 150 models both rear engine mounting bolts are concealed by rubber plugs on the right-hand side of the crankcase. With the aid of an assistant, support the engine unit and withdraw both mounting bolts. The engine can then be lowered from the frame and positioned on the workbench.

5   Dismantling the engine/gearbox unit: preliminaries

1   Before any dismantling work is undertaken, the external surfaces of the unit should be thoroughly cleaned and degreased. This will prevent the contamination of the engine internals, and will also make working a lot easier and cleaner. A high flash-point solvent, such as paraffin can be used, or better still, a proprietary engine degreaser such as Gunk or Jizer. Use old paintbrushes and toothbrushes to work the solvent into the various recesses of the engine castings. Take care to exclude solvent or water from the electrical components and inlet and exhaust ports. The use of petrol as a cleaning medium should be avoided because of the fire risk.

2   When clean and dry, arrange the unit on the workbench, leaving a suitable clear area for working. Gather a selection of small containers and plastic bags so that parts can be grouped together in an easily identifiable manner. Some paper and a pen should be on hand to permit notes to be made and labels attached where necessary. A supply of clean rag is also required.

3   Before commencing work, read through the appropriate section so that some idea of the necessary procedure can be gained. When removing the various engine components it should be noted that great force is seldom required, unless specified. In many cases, a component's reluctance to be removed is indicative of an incorrect approach or removal method. If in any doubt, re-check with the text.

6   Dismantling the engine/gearbox unit: removing the cylinder head, barrel and piston

1   These items can be removed with the engine unit in or out of the frame but in the former case the fuel tank, carburettor and exhaust system must be removed as described in Chapter 2. Additionally, it will be necessary to remove the right-hand crankcase cover and gearbox sprocket as described in Section 4 of this Chapter, then slacken and remove the two nuts and washers which secure the cylinder head to the upper engine mounting. Slacken both the engine rear mounting bolts and allow the engine unit to tilt forward; this will gain the required clearance to remove the cylinder head and barrel.

2   Remove the spark plug and the heat dissipator (where fitted) from the top of the cylinder head, and remove the four rubber noise dampers from the cooling fins of the head and barrel. Working in a diagonal sequence slacken each of the cylinder head retaining nuts by a $\frac{1}{4}$ of a turn at a time until all have been loosened. Remove the nuts and washers and lift away the cylinder head. Remove the head gasket and measure its thickness using a micrometer. This information will be needed when purchasing a new gasket since they are available in various thicknesses (see Section 36). Having taken and recorded the measurement, discard the gasket; a new one must be fitted on reas-

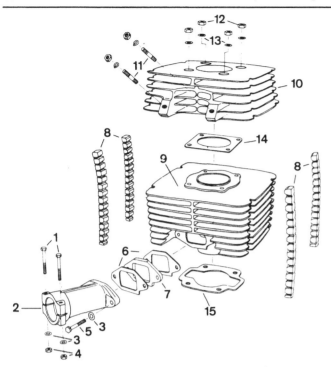

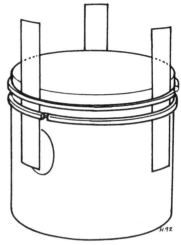

Fig. 1.2 Method of removing and refitting piston rings (Sec 6)

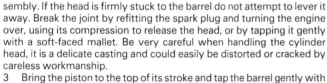

Fig. 1.1 Cylinder head and barrel – typical

1	Bolt – 2 off	9	Cylinder barrel
2	Inlet stub	10	Cylinder head
3	Spring washer – 4 off	11	Stud – 2 off
4	Nut – 2 off	12	Nut – 4 off
5	Bolt – 2 off	13	Washer – 4 off
6	Gasket – 2 off	14	Head gasket
7	Insulator	15	Base gasket
8	Noise damper – 4 off		

reassembly.

5   If the gudgeon pin is a tight fit in the piston, soak a rag in boiling water, wring it out and wrap it around the piston; the heat will expand the piston sufficiently to release its grip on the pin. If necessary, the pin may be tapped out using a hammer and drift, but take care to support the piston and connecting rod in the process.

6   The piston rings are removed by holding the piston in both hands and gently prising the ring ends apart with the thumb nails until the rings can be lifted out of their grooves and onto the piston lands, one side at a time. The rings can then be slipped off the piston and put to one side for cleaning and examination. If the rings are stuck in their grooves by excessive carbon deposits use three strips of metal to remove them, as shown in the accompanying illustration. Be careful as the rings are brittle and will break easily if overstressed.

7   Dismantling the engine/gearbox unit: removing the generator

1   The alternator can be removed with the engine in or out of the frame but in the former case the right-hand crankcase cover must be removed and the wiring disconnected. Refer to Section 4 for further information.

2   Disconnect the wires from the carbon brush terminals, slacken the two screws which retain the brushholder to the stator housing and remove it from the generator.

3   Slacken and remove the three stator housing retaining screws and pull the stator housing off the end of the crankshaft. Slacken the rotor retaining bolt and remove the bolt together with the spring washer and cam. There should be no need to lock the crankshaft; a sharp tap on the end of a long ring spanner should be quite sufficient to release the bolt.

4   To extract the rotor, obtain a metric bolt of 10 mm thread size and at least 100 mm (4 in) long. Screw this into the thread cut in the outer end of the rotor until it tightens against the crankshaft end, then tap smartly on the bolt head to release the joint. Withdraw the rotor and remove the

sembly. If the head is firmly stuck to the barrel do not attempt to lever it away. Break the joint by refitting the spark plug and turning the engine over, using its compression to release the head, or by tapping it gently with a soft-faced mallet. Be very careful when handling the cylinder head, it is a delicate casting and could easily be distorted or cracked by careless workmanship.

3   Bring the piston to the top of its stroke and tap the barrel gently with a soft-faced mallet to break the seal, then lift the barrel just enough to expose the bottom of the piston skirt. Pack a wad of clean rag into the crankcase mouth to prevent dirt or debris falling in, then lift the barrel away. Remove the base gasket from the crankcase mouth.

4   Use a pair of pointed-nosed pliers to remove one of the gudgeon pin circlips then press out the gudgeon pin far enough to clear the connecting rod and withdraw the piston. Push out the small-end bearing. Discard the used circlips and base gasket and obtain new ones for the

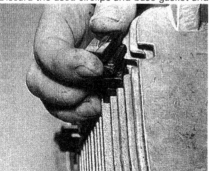

6.2 Do not forget to remove noise dampers before removing cylinder head

7.2a Disconnect brushholder wires ...

7.2b ... and remove it from stator housing

Woodruff key or pin (as applicable) from the crankshaft and store it with the rotor for safekeeping.

5   *Do not attempt to remove the rotor by any other means and never strike it or use excessive force during removal.* If it proves stubborn, take the machine to an authorized MZ dealer for the rotor to be removed using service tool 02-MW 39-4.

8 Dismantling the engine/gearbox unit: removing the left-hand crankcase cover

1   This can be removed with the engine unit either in or out of the frame, but in the former case it will be necessary to first drain the transmission oil as described in Routine maintenance, and to disconnect the oil pipes and operating cable from the oil pump (where fitted). On 250, 251 and 300 models it will also be necessary to remove the tachometer drive housing or cover (as applicable) and disconnect the clutch cable. Refer to Section 4 of this Chapter for further information.

2   Slacken the gearchange lever pinch bolt and pull the lever off the splined shaft. It is advisable to mark the shaft and lever so that the lever can be fitted in its original position on reassembly. On 125 and 150 models repeat the process for the kickstart lever. On 250, 251 and 300 models slacken the nut or tachometer drive gear (as applicable) from the left-hand end of the crankshaft and remove it along with its flat washer. To do this it will be necessary to prevent the crankshaft rotating using one of the methods described in paragraph 3 of the following Section.

3   On all models, remove the left-hand crankcase cover retaining screws, then tap the cover gently with a soft-faced mallet to break the joint, and remove it from the engine unit. Be prepared to catch any surplus oil which may be released as the casing is removed. On 125 and 150 models remove the thrust washer from the kickstart shaft and store it with the casing for safekeeping. Note that on 250, 251 and 300 models, if removal of the kickstart lever or shaft is required, refer to Section 10.

9 Dismantling the engine/gearbox unit: removing the clutch and primary drive components

1   These can be removed with the engine unit either in or out of the frame. Remove the left-hand crankcase cover as described in Section 8 of this Chapter then follow the procedure given under the relevant sub-heading. Note that if on 125 and 150 models it is wished to check the primary chain for wear, this can be done with the chain in situ as described in Section 24.

125 and 150 models

2   Before the clutch is removed there is one consideration which should be made. If the work is being carried out with the engine in the frame the bolt which secures the primary drive sprocket must be slackened at this stage if it is wished to remove the sprocket, primary drive chain and clutch drum. To prevent crankshaft rotation as the bolt is removed, select top gear and apply the rear brake hard. With the rear wheel in firm contact with the ground the engine can be locked through the gear train whilst the bolt is slackened. Alternatively, if engine has been removed from the frame and the top end of the engine has also been removed, it is possible to lock the crankshaft by passing a close fitting metal bar through the connecting rod small-end eye and resting it on wooden blocks placed across the crankcase mouth.

3   Straighten the three clutch bolt locking tabs and slacken and re-move the bolts along with the tab washers and triangular shims (if fitted). Remove the support plate, clutch spring and the clutch pressure plate. Withdraw all the friction and plain plates noting the taper on the inner plain plate. Withdraw the mushroom-headed pushrod from the centre of the input shaft.

4   Before the clutch centre nut can be removed, some method of preventing the centre rotating must be devised. If the work is being carried out with the engine in the frame, the easiest way of achieving this is to select top gear and apply the rear brake hard. If the work is being carried out with the engine/gearbox unit removed from the frame, it will be necessary to construct a holding tool, as shown in the accompanying illustration. This tool was made up from 1 in x ⅛ in mild steel strip, the edges of the angled jaws being ground to fit snugly in the clutch centre splines. An assistant will be required to hold the clutch centre with the home-made tool while the nut is slackened. *Take care not to allow the tool to slip or the soft alloy splines of the clutch centre will be damaged.*

5   With the clutch centre held, slacken the clutch centre nut, noting that it has a **left-hand thread** and is slackened in a **clockwise** direction. Remove the nut and tab washer and pull the clutch centre off the mainshaft, followed by the plain thrust washer. The clutch outer drum, primary drive chain and primary drive sprocket must be removed as a single unit after withdrawing the sprocket as follows.

6   To remove the primary drive sprocket from the crankshaft an extractor, MZ reference number 12 MV 32-4, is required. If this extractor is available, remove the sprocket retaining bolt, unscrew the centre bolt of the extractor then screw the extractor tightly into the sprocket. Ensure the extractor is screwed fully into the sprocket then tighten the extractor centre bolt against the crankshaft end and tap smartly on the end of the centre bolt to break the joint between the sprocket and shaft. If necessary repeat the process until the sprocket is released from the crankshaft end. If the tool is not available a legged puller can be used as described above, provided that the sprocket retaining bolt is refitted to protect the crankshaft end. *Do not apply excessive force if a legged puller is used.* If the sprocket proves stubborn take the machine to an autho-rized MZ dealer for the task to be performed with the correct service tool. When the sprocket is released, withdraw it as a single unit along with the chain and clutch drum. Once the clutch drum has been removed slide the clutch drum collar off the input shaft followed by the shim(s) (where fitted) and thrust washer.

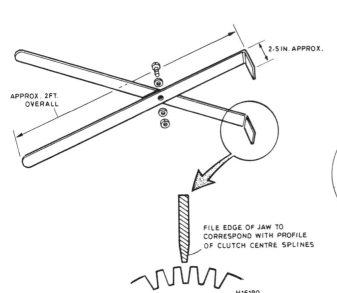

Fig. 1.3 Fabricated clutch holding tool – 125 and 150 models (Sec 9)

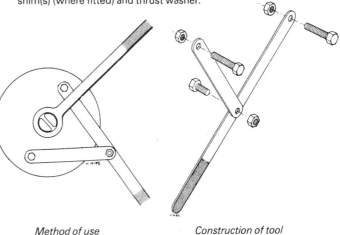

Method of use   *Construction of tool*

Fig. 1.4 Fabricated primary driven gear holding tool – 250, 251 and 300 models (Sec 9)

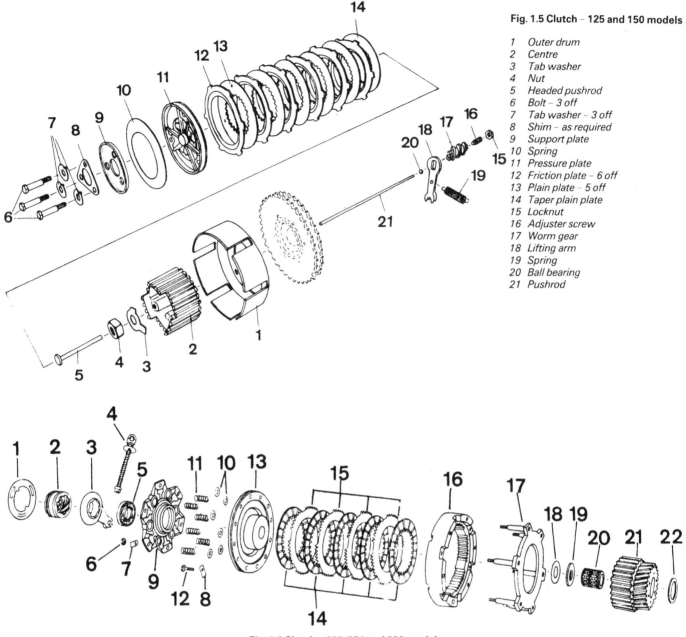

Fig. 1.5 Clutch – 125 and 150 models

1   Outer drum
2   Centre
3   Tab washer
4   Nut
5   Headed pushrod
6   Bolt – 3 off
7   Tab washer – 3 off
8   Shim – as required
9   Support plate
10 Spring
11 Pressure plate
12 Friction plate – 6 off
13 Plain plate – 5 off
14 Taper plain plate
15 Locknut
16 Adjuster screw
17 Worm gear
18 Lifting arm
19 Spring
20 Ball bearing
21 Pushrod

Fig. 1.6 Clutch – 250, 251 and 300 models

1 Adjusting plate	7 Tab washer – 6 off	13 Top plate	19 Thrust washer
2 External worm gear	8 Tab washer – 6 off	14 Friction plate – 5 off	20 Needle roller bearing
3 Internal worm gear	9 Pressure flange	15 Plain plate – 4 off	21 Primary drive gear
4 Connecting rod	10 Washer – 6 off	16 Clutch body	22 Thrust washer
5 Bearing	11 Spring – 6 off	17 Pressure plate	
6 Nut – 6 off	12 Bolt – 6 off	18 Spring washer	

250, 251 and 300 models

7   If the work is being carried out with the engine out of the frame, there is one consideration which should be made. If the large primary driven gear is to be removed, its retaining nut should be slackened at this stage, before the clutch is removed. The primary driven gear can then be locked up through the crankshaft as described in paragraph 3. If the work is being carried out with the engine in the frame the gear can be locked up through the transmission and held on the rear brake, also as described in paragraph 3. Alternatively a home-made holding tool, like that shown in the accompanying illustration and photo, can be made to retain the large primary driven gear. This was made up with some steel strip and a few bolts and need not be very elaborate.

8   The clutch assembly is removed from the end of the crankshaft in the same way as the primary drive sprocket is removed on 125 and 150

models, requiring the use of a special MZ extractor as described in paragraph 6. No other method of removing the clutch is recommended as the clutch is likely to be a very tight fit on the crankshaft and alternative removal methods could well cause damage to the clutch assembly. Therefore, if this extractor is not available the machine must be taken to an authorized MZ dealer for the clutch to be removed.

9   Once the clutch has been removed, slide the spring and thrust washers off of the crankshaft followed by the small primary drive gear. Then remove the needle roller bearing and second thrust washer. If necessary the clutch can be dismantled as described in paragraph 12.

10 The large primary driven gear can be removed using a legged puller, ensuring that the gear retaining nut is fitted to protect the input shaft end whilst the gear is removed. Once the gear is free of the shaft remove the nut and tab washer and lift it away from the machine.

9.7 On 250, 251 and 300 models primary driven gear can be retained using home-made holding tool whilst nut is slackened ...

9.8 ... and clutch must be removed using an extractor

11   Before dismantling the clutch, mark the outer drum, pressure plate and flange so that all these components can be returned to their original positions on reassembly. This is necessary to ensure the clutch remains perfectly balanced and operates smoothly.
12   To dismantle the clutch straighten the locking tabs of the six clutch assembly nuts. Place the assembly in a vice equipped with soft jaws and tighten the vice gently to compress the assembly slightly and remove the spring pressure from the pressure plate and clutch nuts. Then slacken and remove the six nuts and tab washers and gently undo the vice. Lift off the pressure flange and remove the springs and washers. Withdraw the pressure plate from the rear of the clutch body and withdraw the plain and friction plates from inside the body.

10   Dismantling the engine/gearbox unit: removing the kickstart mechanism

1   This can be removed with the engine unit either in or out of the frame. Remove the left-hand crankcase cover as described in Section 8 of this Chapter then follow the procedure given under the relevant sub-heading.

125 and 150 models
2   To remove the kickstart mechanism it is first necessary to remove the clutch assembly as described in Section 9.
3   Temporarily refit the kickstart lever to the shaft, grasp it firmly and carefully withdraw the shaft until the quadrant clears the crankcase

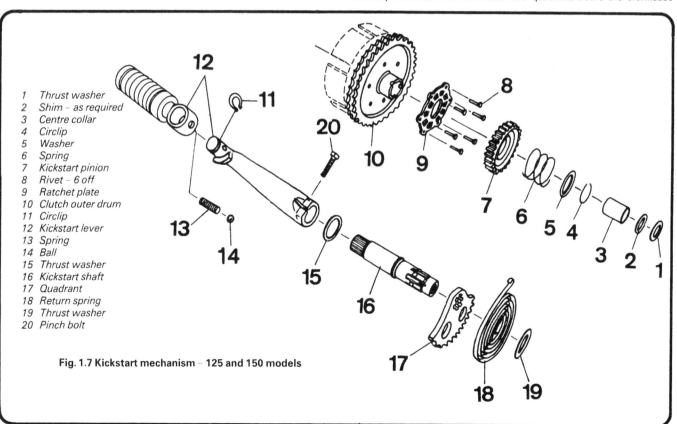

1   Thrust washer
2   Shim – as required
3   Centre collar
4   Circlip
5   Washer
6   Spring
7   Kickstart pinion
8   Rivet – 6 off
9   Ratchet plate
10   Clutch outer drum
11   Circlip
12   Kickstart lever
13   Spring
14   Ball
15   Thrust washer
16   Kickstart shaft
17   Quadrant
18   Return spring
19   Thrust washer
20   Pinch bolt

Fig. 1.7 Kickstart mechanism – 125 and 150 models

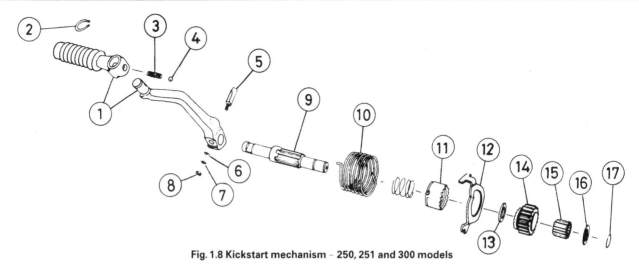

Fig. 1.8 Kickstart mechanism – 250, 251 and 300 models

1	Kickstart lever	6	Washer	11	Ratchet	16	Thrust washer
2	Circlip	7	Spring washer	12	Camplate	17	Circlip
3	Spring	8	Nut	13	Thrust washer		
4	Ball	9	Kickstart shaft	14	Pinion		
5	Cotter pin	10	Return spring	15	Needle roller – 24 off		

10.4a On 125 and 150 models, if necessary, remove circlip ...

10.4b ... thrust washer and spring from back of clutch drum ...

10.4c ... and lift off kickstart pinion

stop. Allow the shaft to rotate slowly until spring pressure is released, then withdraw the shaft along with the quadrant and return spring. Remove the thrust washer fitted behind the kickstart shaft return spring.

4   The kickstart pinion (ratchet) is situated on the back of the clutch drum. Remove the circlip which retains the pinion and withdraw the thrust washer, spring and pinion.

250, 251 and 300 models
5   Remove the nut and washers from the kickstart lever cotter pin and gently tap the cotter pin out of position using a hammer and suitable drift. Take care during this operation as the kickstart return spring pressure will be released when the pin is removed. Then pull the lever off the shaft and remove the shaft assembly from the casing.

6   To dismantle the kickstart shaft, start by removing the circlip from the end of the shaft and sliding off the thrust washer. Before removing the pinion some provision must be made to catch the needle rollers which the pinion runs on. Then carefully remove the pinion, catching all the rollers, followed by a second thrust washer and the cam plate. Before removing the ratchet mark the pinion in some way to use as a reference on reassembly, then slide off the ratchet and spring.

---

11   Dismantling the engine/gearbox unit: removing the tachometer drive – 125 and 150 models

---

1   This can be removed with the engine unit either in or out of the

frame. Firstly remove the left-hand crankcase cover and the clutch assembly as described in Sections 8 and 9 of this Chapter.

2   Remove the circlip which retains the tachometer drive gear then slide off the thrust washer and the gear itself.

3   If required, the driven gear shaft can be pulled out of the crankcase and removed. Straighten the locking tab of the driven gear bush retaining bolt and remove the bolt and locking tab. The bush can then be pulled out of the crankcase followed by the driven gear. Note the thrust washers which are positioned on each end of the driven gear shaft.

---

12   Dismantling the engine/gearbox unit: separating the crankcase halves

---

1   The crankcase halves can only be separated after the engine/gearbox unit has been removed from the frame and all the preliminary dismantling in Sections 6 – 11 has been carried out.

2   **Note**: MZ recommend that their service tools are used to separate the crankcase halves and to remove and refit the crankshaft to avoid the risk of damaging or distorting the crankcase castings and crankshaft assembly. The methods given in this Manual avoid the use of these tools but require extreme care to avoid damage. Owners should note that excessive force will not be required at any point; if a component proves stubborn take the engine to an authorized MZ dealer for the work to be carried out with the correct service tools.

3   On 125 and 150 models, slacken the screws which retain the three

12.3a Slacken all oil seal cap retaining screws ...

12.3b ... and remove cap ...

12.3c ... along with any shims which are fitted behind it

12.3d On 125 and 150 models remove gearchange detent spring from crankcase

12.4a On 250, 251 and 300 models remove detent arm and spring assembly ...

12.4b ... and prise out three rubber plugs which conceal crankcase screws

12.5a Remove neutral detent bolt and spring ...

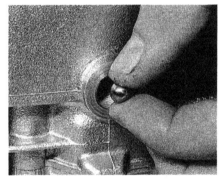
12.5b ... along with the ball bearing

12.8 Remove rubber separating disc for safekeeping

oil seal caps and remove the caps and shims (if fitted) from the engine noting exactly where each shim is positioned. Unhook the gearchange mechanism detent spring, situated just above the gearchange shaft, and remove it from the crankcase.

4   On 250, 251 and 300 models, unhook the detent arm spring from the left-hand side of the crankcase and remove it along with the detent arm. Remove the circlip which retains the output shaft oil guide cap and plate and withdraw them from the crankcase. On the right-hand side, slacken the three screws which retain the output shaft oil seal cap and remove it along with any shims which may be fitted behind it. Use a screwdriver to prise out the three black rubber plugs which conceal crankcase screws. Slacken and remove the 8 mm bolt, situated in front of the crankcase mouth, which secures the crankcase halves together.

5   On all models slacken all the crankcase screws (11 on 125 and 150 models and 14 on 250, 251 and 300 models) progressively and in a diagonal sequence. It is a good idea to make up a cardboard template into which the screws can be inserted as they are removed, thus preventing their loss and ensuring that they are refitted in their original

positions. Unscrew the neutral detent bolt from the bottom of the crankcase and remove it along with its spring and ball bearing.

6   Place two clean wooden blocks on the workbench and support the engine/gearbox unit on them so that the crankcase right-hand side is uppermost. The crankcase right-hand half is to be drawn off, leaving the crankshaft and gearbox components in the left-hand half.

7   Using only a soft-faced mallet, tap gently on the exposed ends of the crankshaft and gearbox shafts and all around the joint area of the two crankcase halves until initial separation is achieved. Lift off the right-hand half, ensuring that it remains absolutely square so that the bearings do not stick on their respective shafts; tap gently on the shaft ends to assist removal. *Never attempt to lever the crankcases apart; this will damage the sealing faces and may distort the castings.*

8   When the casing is removed, check that there are no loose components such as dowels or thrust washers sticking to it which might drop clear and be lost. Any such components should be refitted in their correct places. Remove the rubber separating disc which is situated between the crankshaft and gearbox shafts in the left-hand casting and

store it in a safe place. Unless the locating dowels are firmly fixed in position they too should be removed for safekeeping.

13   Dismantling the engine/gearbox unit: removing the crankshaft and gearbox components

1   When the crankcase halves have been separated, the crankshaft and gearbox components may be removed. Note that the crankshaft can be removed with the gearbox components still fitted to the crankcase half and it is therefore recommended that the gearbox components are not unnecessarily disturbed.

125 and 150 models
2   To remove the gearbox components first withdraw the selector fork shaft from the casing. Then remove the selector forks taking great care to note exactly where each one is fitted and in what way it is fitted. It is useful to degrease each fork as it is withdrawn and to mark it with a spirit-based felt pen as a guide to correct assembly, but note that the forks fitted to the machines used in the photographs were identified by numbers and model codes cast in each fork. If this is the case on the machine being worked on, these marks can be used providing notes are made of their positions. As each fork is withdrawn replace it on the shaft to prevent any confusion on reassembly.
3   Unhook the gearchange shaft claw from selector drum and pull the selector drum out of the casting. The gearchange shaft can then also be removed.

250, 251 and 300 models
4   Disengage the gearchange shaft claw from the selector drum and remove the shaft from the casing. Withdraw the selector fork shaft taking care not to lose the two thrust washers fitted to each end of the shaft.
5   Referring to paragraph 3 of this Section for details of identification marks, withdraw the upper (right-hand) selector fork from the output shaft and the centre selector fork from the input shaft. Disengage the lower (left-hand) selector fork from the selector drum and its pinion so that the guide pin is clear of the drum and there is sufficient space for the drum to be withdrawn. This is done by pulling the drum out of the casing. Once the drum has been removed the lower selector fork can be removed from the casing.

All models
6   The gearbox shafts should be removed simultaneously, noting that it may be necessary to tap the ends of the shafts lightly with a soft-faced mallet to free them from their bearings.
7   Before attempting to remove the crankshaft, refer to the warning note given in Section 12, paragraph 2.
8   To remove the crankshaft support the crankcase half on two wooden blocks placed as close as possible around the crankshaft to provide maximum support for the crankcase. The blocks should be of a size to hold the crankcase far enough from the workbench to allow the crankshaft to be removed. Refit the nut or bolt to the left-hand end of the crankshaft to protect the threads of the shaft, noting that on 250, 251 and 300 models fitted with a tachometer, it is recommended that a spare 14 mm nut is used for this purpose and not the tachometer drive gear. Using a soft-faced mallet with one hand, and supporting the crankshaft with the other, carefully tap the crankshaft out of the casting. Do not use excessive force and do not allow the crankshaft to drop out of the casting onto the workbench.

14   Dismantling the engine/gearbox unit: removing the oil seals and bearings

1   All oil seals with the exception of the crankshaft main bearing seals on 250, 251 and 300 models, can be renewed without separating the crankcases. They are fitted into caps screwed to the crankcase and are easily removed as described in Section 12. When the caps are removed take care not to lose any shims which may be positioned behind them. If the main bearing oil seals require renewal on 250, 251 and 300 models it will be necessary to first separate the crankcases, remove the crank-

shaft, and prise out the circlips which retain the seals.
2   Oil seals are easily damaged when disturbed and, thus, should be renewed as a matter of course during overhaul. Prise them out of position using the flat of a screwdriver whilst taking care not to damage the alloy seal housing; note which way round the oil seal is fitted.
3   The crankshaft and gearbox bearings are a press fit in their respective crankcase locations. To remove the bearing, the crankcase casting must be heated so that it expands and releases its grip on the bearing, which can then be drifted or pulled out. Before the casting is heated, remove any circlips, shims or seals which restrict access to the bearing outer race.
4   To prevent casting distortion, it must be heated evenly to a temperature of about 100°C (212°F) by placing it in an oven; if an oven is not available, place the casting in a suitable container and carefully pour boiling water over it until it is submerged. Note that if heat is used in this way, care must be taken first to remove any components which would be damaged by excess heat, such as the neutral switch.
5   Taking great care to prevent personal injury when handling the heated components, lay the casting on a clean surface and tap out the bearing using a hammer and suitable drift. If the bearing is to be re-used apply the drift only to the bearing outer race, where accessible, to avoid damaging the bearing. In some cases it may be necessary to pressure the bearing inner race; in such cases closely inspect the bearing for signs of damage before using it again. When drifting a bearing from its housing it must be kept square to the housing to prevent it from jamming with the resulting risk of damage. Where possible, use a tubular drift such as a socket spanner which bears only on the outer race of the bearing; if this is not possible, tap evenly around the outer race to achieve the same result.

15   Examination and renovation: general

1   Before any component is examined, it must be cleaned thoroughly. Being careful not to mark or damage the item in question, use a blunt-edged scraper (an old kitchen knife or a broken plastic ruler can be very useful) to remove any caked-on deposits of dirt or oil, followed by a good scrub with a soft wire brush (a brass wire brush of the type sold for cleaning suede shoes is best, with an assortment of bottle-cleaning brushes for ports etc). Take care not to remove any paint code marks from internal components.
2   Soak the component in a solvent to remove the bulk of the remaining dirt or oil. If one of the proprietary engine degreasers (such as Gunk or Jizer) is not available, a high flash-point solvent such as paraffin should be used. The use of petrol as a cleaning agent cannot be recommended because of the fire risk. With all of the above cleaning agents take great care to prevent any drops getting into the eyes and try to avoid prolonged skin contact. To finish off the cleaning procedure wash each component in hot soapy water (as hot as your hands can bear); this will remove a surprising amount of dirt on its own and the residual heat usually dries the component very effectively.
3   Make sure all traces of old gaskets have been removed and that the mating surfaces are clean and undamaged. Great care should be taken when removing old gasket compound not to damage the mating surface. Most gasket compounds can be softened using a suitable solvent such as methylated spirits, acetone or cellulose thinner. The type of solvent required will depend on the type of compound used. Gasket compound of the non-hardening type can be removed using a soft brass-wire brush of the type used for cleaning suede shoes. A considerable amount of scrubbing can take place without fear of harming the mating surfaces. Some difficulty may be encountered when attempting to remove gaskets of the self-vulcanising type, the use of which is becoming widespread, particularly as cylinder head and base gaskets. The gasket should be pared from the mating surface using a scalpel or a small chisel with a finely honed edge. Do not, however, resort to scraping with a sharp instrument unless necessary.
4   If there is the slightest doubt about the lubrication system, for example if a fault appears to have been caused by a failure of the oil supply, all components should be dismantled so that the oilways can be checked and cleared of any possible obstructions. Always use clean, lint-free rag for cleaning and drying components to prevent the risk of small particles obstructing oilways.
5   Examine each part carefully to determine the extent of wear, checking with the tolerance figures listed in the Specifications section of

this chapter. If there is any doubt about the condition of a particular component, play safe and renew it.

6 Various instruments for measuring wear are required, including an internal and external micrometer or vernier gauge, and a set of standard feeler gauges. Additionally, although not absolutely necessary, a dial gauge and mounting bracket are invaluable for accurate measurement of endfloat, and play between components of very low diameter bores – where a micrometer cannot reach. After some experience has been gained, the state of wear of many components can be determined visually, or by feel, and a decision on their suitability for re-use made without resorting to direct measurement.

16 Examination and renovation: engine cases and covers

1 Small cracks or holes in aluminium castings may be repaired with an epoxy resin adhesive such as Araldite, as a temporary measure. Permanent repairs can only be effected by argon-arc welding, and only a specialist in this process is in a position to advise on the economics or practicability of such a repair.

2 Damaged threads can be economically reclaimed using a diamond-section wire insert, of the Helicoil type, which is easily fitted after drilling and re-tapping the affected thread. Most motorcycle dealers and small engineering firms offer a service of this kind.

3 Sheared studs or screws can usually be removed with screw extractors, which consist of tapered, left-handed thread screws, of very hard steel. These are inserted by screwing anticlockwise into a pre-drilled hole in the stud, and usually succeed in dislodging the most stubborn stud or screw. If a problem arises which seems beyond your scope, it is worth consulting a professional engineering firm before condemning an otherwise sound casing. Many of these firms advertise regularly in the motorcycle press.

4 If the gasket or other mating surfaces are marked or damaged in any way they can be reclaimed by rubbing them on a sheet of fine abrasive paper laid on an absolutely flat surface such as a sheet of glass. Start with 200 grade paper and finish with 400 grade and oil. Use a gentle figure-of-eight pattern, maintaining light but even pressure on the casting. Note that if large amounts of material are to be removed, advice should be sought as to the viability of re-using the casting in question; the internal clearances are minimal in many cases between the rotating or moving components and the castings. Stop work as soon as the entire mating surface is polished by the action of the paper.

5 Note that the mating surface may become distorted outwards around the mounting screw holes, usually because these have been grossly overtightened. In this event, use a large drill bit or countersink to very lightly skim the raised lip from around the screw hole, then clean up the whole surface as described above.

6 Finally, check that all screw or bolt tapped holes are clean down to the bottom of each hole; serious damage can be caused by forcing a screw or bolt down a dirty thread and against an incorrect stop caused by the presence of dirt, oil, swarf or blobs of old jointing compound. At the very least the component concerned will be incorrectly fastened, at worst the casing could be cracked. The simplest way of cleaning holes is to use a length of welding rod or similar to dig out any embedded foreign matter, then to give each hole a squirt of contact cleaner or similar solvent applied from an aerosol via the long extension tube usually supplied. Be careful to wear eye protection while doing this; the amount of dirt and debris that can be ejected from each hole is surprising.

17 Examination and renovation: bearings and oil seals

1 Ball bearings should be washed thoroughly to remove all traces of oil then tested as follows. Hold the outer race firmly and attempt to move the inner race up and down, then from side to side. Examine bearing balls, tracks and cages looking for signs of pitting or other damage. Finally spin the bearing and check that it rotates smoothly and with no sign of notchiness. If any free play, roughness or other damage is found the bearing must be renewed.

2 Roller bearings are checked in much the same way, except that free play can only be checked in the up and down direction with the components temporarily assembled. Remember that if a roller bearing fails it may well mean having to renew, as well as the bearing, one or two other components which form its inner and outer races. If in any doubt about the condition of a roller bearing, renew it.

3 Do not waste time checking oil seals. Discard all seals and O-rings disturbed during dismantling work and fit new ones on reassembly. Considering their habit of leaking once disturbed, and the amount of time and trouble necessary to renew them, they are relatively cheap if renewed as a matter of course whenever they are disturbed.

18 Examination and renovation: cylinder head

1 Check that the cylinder head fins are not clogged with oil or road dirt, otherwise the engine will overheat. If necessary, use a degreasing agent to clean between the fins. Check that no cracks are evident, especially in the vicinity of the spark plug or bolt holes.

2 Remove all traces of carbon from the cylinder head, using a blunt-ended scraper. Finish by polishing with metal polish, to give a smooth shiny surface. This will aid gas flow and will also prevent carbon from adhering so firmly in the future.

3 Check the condition of the threads in the spark plug hole. If the threads are worn or stretched as the result of overtightening the plug, they can be reclaimed by a Helicoil thread insert. Most dealers have the means of providing this cheap but effective repair.

4 If the cylinder head joint has shown signs of leakage, check whether the cylinder head is distorted by laying it on a sheet of plate glass and measuring distortion with feeler gauges. Note that a maximum warpage figure is not given by the manufacturer, although this test should give some indication of the extent of distortion. Severe distortion will necessitate renewal of the head but if distortion is only slight it can be removed by rubbing the mating surface of the head in a slow circular motion against emery paper placed on plate glass (see Section 16). Do not forget to check the corresponding surface of the cylinder barrel for distortion.

5 Note that most cases of cylinder head distortion can be traced to unequal tension of the cylinder head retaining nuts or to tightening them in the incorrect sequence.

19 Examination and renovation: cylinder barrel

1 The usual indication of cylinder barrel and piston is piston slap, a metallic rattle that occurs when there is little or no load on the engine.

2 Clean all dirt from between the cooling fins of the barrel. Carefully remove the ring of carbon from the bore mouth so that bore wear can be accurately assessed and check the barrel/cylinder head mating surface for warpage as described in the previous Section. Clean all carbon from the exhaust port and all traces of old gasket from the cylinder base.

3 Examine the bore for scoring or other damage, particularly if broken rings are found. Damage of this kind will necessitate reboring and a new piston regardless of the amount of wear. A satisfactory seal cannot be obtained if the bore is not perfectly finished.

4 There will probably be a lip at the uppermost end of the cylinder bore which marks the limit of travel of the top piston ring. The depth of this lip will give some indication of the amount of bore wear that has taken place even though the amount of wear is not evenly distributed.

5 The most accurate way of measuring bore wear is by the use of a cylinder bore DTI (Dial Test Indicator) or a bore micrometer. Measure at the top (just below the wear ridge), middle and bottom of the bore, both in line with the gudgeon pin axis and at 90° to it, avoiding the port areas, taking six measurements in all. Measure the piston as described in the following Section and subtract the piston diameter from the maximum bore figure obtained. If the difference calculated exceeds the service limit given in the Specifications the bore is excessively worn, assuming the piston is not worn (see Section 20).

6 Alternatively, insert the piston in the bore in its correct position with the skirt just below the wear ridge, ie at the point of maximum wear. Measure the gap using feeler gauges. If the piston/cylinder clearance exceeds the service limit the barrel must be rebored. It must be stressed that this can only be used as a guide to the degree of bore wear, unless the piston is known to be unworn; have the barrel measured accurately by an authorized MZ dealer or similar expert before a rebore is undertaken.

7 After a rebore has been carried out, the reborer should hone the bore lightly to provide a fine cross-hatched surface so that the new piston and rings can bed in correctly. Also the edges of the ports should be chamfered first with a scraper, then with fine emery, to prevent the rings from catching on them and breaking.

8   If a new piston and/or rings are to be run in a part-worn bore the surface should be prepared first by glaze-busting. This involves the use of a special honing attachment with (usually) an electric drill to provide a surface similar to that described above. Most motorcycle dealers have such equipment and will be able to carry out the necessary work for a small charge.

20   Examination and renovation: piston and rings

1   Disregard the existing piston and rings if a rebore is necessary; they will be replaced with oversize items. Note also that it is considered a worthwhile expense by many mechanics to renew the piston rings as a matter of course, regardless of their apparent condition.

2   Measure the piston diameter at right angles to the gudgeon pin axis, at a point approximately 15 mm (0.6 in) above the base of the skirt. Since the manufacturer does not give any service limits, the degree of wear can be assessed only by direct comparison with a new component. Ensure that the new piston is of the same size (standard or oversize) as the original, the actual diameter being stamped on the piston crown.

3   Piston wear usually occurs at the skirt, especially on the forward face, and takes the form of vertical score marks. Reject any piston which is badly scored or has been blackened as the result of the blow-by of gas. Slight scoring of the piston can be removed by careful use of a fine swiss file. Use chalk to prevent the file teeth clogging and the subsequent risk of scoring. If the ring locating pegs are loose or worn, renew the piston.

4   The gudgeon pin should be a firm press fit in the piston. Check for scoring on the bearing surfaces of each part and where damage or wear is found, renew the affected part. The circlip retaining grooves must also be undamaged; renew the piston rather than risk damage to the bore through a circlip becoming detached. *Discard the circlips; they should never be re-used.*

5   Any build-up of carbon in the ring grooves can be removed using a section of broken piston ring, the end of which has been ground to a chisel edge. Use feeler gauges to measure the width of each piston ring groove. Renew the piston if any groove has worn to beyond the service limit given in the Specifications. Also measure the height of the piston rings using a micrometer. If any ring has worn to or below its service limit, the rings must be renewed as a set.

6   Measure ring wear by inserting each ring into the bore approximately 10 mm (0.4 in) below the top of the bore and measuring the gap

20 6 Measuring piston ring end gap

between the ring ends with a feeler gauge. Use the piston crown to ensure that the ring is located squarely in the bore. If the measurement exceeds that given in the specifications, renew the rings as a set.

7   Reject any rings which show discoloured patches on their mating surfaces; they should be brightly polished from firm contact with the cylinder bore. 8 Do not assume when fitting new rings that their end gaps will be correct; the installed gap must be measured as described above to ensure that it is within the specified tolerances; if the gap is too wide another piston ring set must be obtained (having checked again that the bore is not excessively worn), but if the gap is too narrow it must be widened by the careful use of a fine file.

21   Examination and renovation: crankshaft

1   Big-end failure is characterised by a pronounced knock which will be most noticeable when the engine is worked hard. The usual causes of failure are normal wear, or failure of the lubrication supply. In the case of

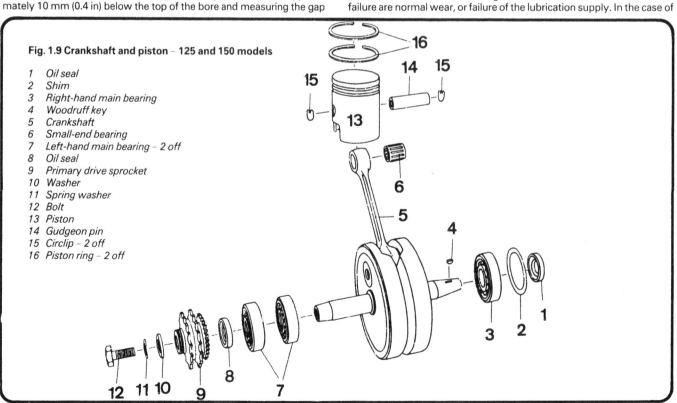

Fig. 1.9 Crankshaft and piston – 125 and 150 models

1   Oil seal
2   Shim
3   Right-hand main bearing
4   Woodruff key
5   Crankshaft
6   Small-end bearing
7   Left-hand main bearing – 2 off
8   Oil seal
9   Primary drive sprocket
10   Washer
11   Spring washer
12   Bolt
13   Piston
14   Gudgeon pin
15   Circlip – 2 off
16   Piston ring – 2 off

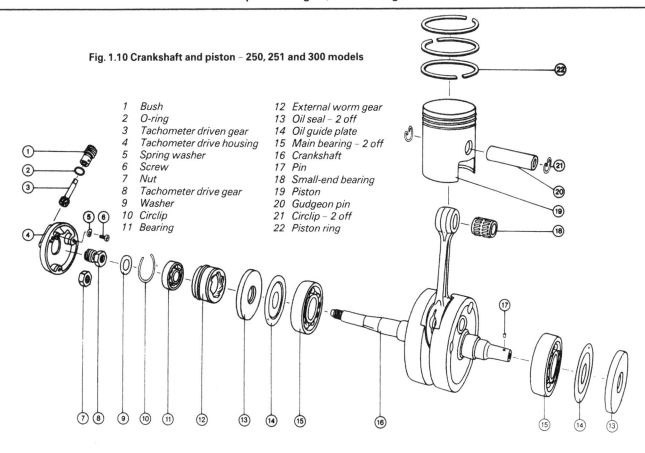

Fig. 1.10 Crankshaft and piston – 250, 251 and 300 models

1 Bush	12 External worm gear
2 O-ring	13 Oil seal – 2 off
3 Tachometer driven gear	14 Oil guide plate
4 Tachometer drive housing	15 Main bearing – 2 off
5 Spring washer	16 Crankshaft
6 Screw	17 Pin
7 Nut	18 Small-end bearing
8 Tachometer drive gear	19 Piston
9 Washer	20 Gudgeon pin
10 Circlip	21 Circlip – 2 off
11 Bearing	22 Piston ring

the latter, the noise will become apparent very suddenly, and will rapidly worsen.

2 Check for wear with the crankshaft set in the TDC (Top Dead Centre) position, by pushing and pulling the connecting rod. No discernible movement will be evident in an unworn bearing, but care must be taken not to confuse endfloat, which is normal, and bearing wear.

3 If the measuring facilities are available, set the crankshaft in V-blocks and check the big-end radial clearance using a dial gauge mounted on a suitable stand. Big-end radial clearance is measured as the amount of up and down movement in the connecting rod. Crank-shaft runout can also be checked with the V-blocks and dial gauge; the gauge pointer being applied to the straight, non-tapered, part of each mainshaft. Big-end axial clearance (endfloat) is measured with feeler gauges of the correct thickness which would be a firm sliding fit between the thrust washer next the big-end eye and the flywheel. If any clearance exceeds the service limits given in the Specifications the crankshaft assembly should be taken to an authorized MZ dealer or similar repair agent for repair or renewal.

4 Push the small-end bearing into the connecting rod eye and push the gudgeon pin through the bearing. Hold the rod steady and feel for movement between it and the gudgeon pin. If movement is felt, renew the pin, bearing or connecting rod as necessary so no movement exists. Renew the bearing if its roller cage is cracked or worn.

5 It should be noted that crankshaft repair work is of a highly special-ised nature and requires the use of equipment and skills not likely to be available to the average private owner. Such work should not be attempted by anyone without this equipment and the skill to use it.

21.3 Measuring big-end axial clearance (endfloat)

22 Examination and renovation: clutch

1 After an extended period of service, the friction plates will have become sufficiently worn to warrant renewal, to avoid subsequent problems with clutch slip. The lining thickness is measured across the friction plate using a vernier caliper. If any plate is worn to or beyond the service limit, the friction plates must be renewed as a set. Note that if new friction plates are fitted, they must be coated with a light film of the

recommended transmission oil before fitting.

2 The plain plates should be free from any signs of blueing, which would indicate that the clutch had overheated in the past. Measure the thickness of each plate and check each plate for distortion by laying it on a flat surface, such as a sheet of plate glass or similar, and measure any detectable gap using feeler gauges. If any of the plates are found to have worn or distorted beyond the service limit all the plain plates should be renewed as a set.

3 The clutch spring(s) may, after a considerable mileage, require renewal. Since the manufacturers do not specify any wear limits, the degree of wear can be assessed only by direct comparison with a new component. On 250, 251 and 300 models the condition of the springs can be judged by measuring their free length. If the free length of any spring is considerably less than the figure given in the Specifications,

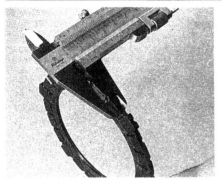

22.1 Measuring clutch friction ...

22.2 ... and plain plate thickness

22.7 Examine clutch lifting mechanism worm gear for wear – 251 shown

the clutch springs should be renewed as a set. On all models, if any doubt remains about the condition of the clutch spring(s), it/they should be renewed.

4   On 125 and 150 models check the condition of the slots in the clutch centre and drum, and on 250, 251 and 300 models in the clutch body and small primary drive gear. In extreme cases, clutch chatter may have caused the tongues of the friction or plain plates to make indentations in the clutch components. These indentations will trap the clutch plates as they are freed and impair clutch action. If the damage is only slight the indentations can be removed by careful work with a file and the burrs from the tongues of the clutch plates removed in a similar manner. More extensive damage will necessitate renewal of the parts concerned.

5   On 125 and 150 models, examine the pushrods for wear and distortion, rolling them on a sheet of plate glass or other flat surface. If bent, the pushrods may be straightened, but if incorrect adjustment or lack of lubrication has caused such a build-up of friction that the hardening on the ends of the pushrods has broken down, the pushrods must be renewed. Such a case will be easy to see due to the heavy blueing and distorted shape which will result.

6   On 250, 251 and 300 models check the pressure flange bearing for signs of wear. There should be no sign of free play between the inner and outer race of the bearing, and the bearing should spin easily with no sign of roughness. If not the bearing must be renewed. On fitting, the new bearing must be staked in place at three points, approximately 120° apart, to ensure that the outer race of the bearing does not spin in the pressure flange.

7   If attention to the clutch lifting mechanism is necessary, remove the lifting arm from the crankcase cover (left-hand on 250, 251 and 300 models and right-hand on 125 and 150 models) and examine the worm gear of the lifting mechanism. If the gear shows signs of wear or damage and fails to operate smoothly it must be renewed. On 125 and 150 models the external worm gear is cast into the crankcase cover so if renewal is required the complete casing must be renewed, whereas on 250, 251 and 300 models the external gear can be driven out of the casing using a suitable hammer and drift and renewed separately. Examine the bearing which is fitted inside the gear for signs of wear or damage, renew it if necessary. The bearing can be drifted out of position once its retaining circlip has been removed.

23 Examination and renovation: kickstart

1   The kickstart mechanism is a robust assembly and should not normally require attention. Apart from obvious defects such as a broken return spring, the kickstart ratchet is the only component likely to cause problems if it becomes worn or damaged.

2   Examine the teeth of the ratchet for signs of wear or damage. If either component shows any sign of wear both the pinion and ratchet should be renewed as a set. On 125 and 150 models the ratchet plate is secured to the back of the clutch drum by six rivets and can be removed and renewed as described in Section 24.

3   The condition of the kickstart return spring and the ratchet spring can be assessed only by direct comparison with a new component. If there is any doubt about the condition of these springs, they should be renewed.

23.2 On 125 and 150 models kickstart ratchet plate is riveted to clutch drum

4   Examine the teeth of the kickstart quadrant and pinion for wear or damage. In view of the fact that these components are not subject to constant use, a significant amount of wear or damage is unlikely to be found. However, if damage is discovered both the quadrant and pinion should be renewed as a set.

5   On 250, 251 and 300 models examine the camplate for any sign of damage or distortion and renew if necessary.

24 Examination and renovation: primary drive

125 and 150 models

1   To check the primary drive chain for wear it is necessary to temporarily install the drive sprocket, chain and clutch drum on the engine unit. The chain wear can then be assessed by measuring the total up and down movement of the chain. If this distance exceeds 10 mm (0.4 in) the chain be considered worn out and must be renewed.

2   Examine the teeth of both the drive sprocket and clutch drum sprocket for signs of wear or damage such as chipped teeth. If either sprocket is damaged, both sprockets and the chain should be renewed as a set.

3   The clutch drum sprocket is available separately and can be renewed individually. To remove the sprocket, drill out the six rivets which secure the sprocket and kickstart ratchet to the clutch drum and separate the three components. On refitting ensure the three components are securely riveted together with no sign of freeplay between the sprocket and clutch drum.

250, 251 and 300 models

4   Examine the teeth of the primary drive and driven gears for signs of

wear or damage, such as chipped teeth. If either gear is worn or damaged both gears must be renewed as a matched pair.

5 If the necessary measuring equipment is available, primary drive wear can be assessed by measuring the amount of backlash present. If the backlash exceeds the service limit, both gears must be renewed as a matched pair.

25 Examination and renovation: gearbox components

1 Give the gearbox components a close visual inspection for signs of wear or damage such as broken or chipped teeth, worn dogs, damaged or worn splines and bent selectors. Renew any parts that are found to be unserviceable because they cannot be reclaimed in a satisfactory manner.

2 The gearbox shafts are unlikely to sustain damage unless the lubricating oil has been run low or the engine has seized and placed an unusually high loading on the gearbox. Check the surfaces of the shaft, especially where a pinion turns on it, and renew the shaft if it is scored or has picked up.

3 The gearbox shafts can be checked for trueness by removing all components from the shaft, as described in Section 26, and setting the shaft up in V-blocks. Slowly turn the shaft whilst measuring any runout with a dial gauge. Since the manufacturer does not give any service limits, the degree of wear can only be assessed by direct comparison with a new item, although if a significant amount of runout is measured the shaft must be renewed.

4 Examine all the pinions for signs of wear or damage. All the gear dogs should have a slight undercut (3°). If this has been reduced by wear, the machine will jump out of gear, necessitating renewal of the worn pinions. Always renew meshing gears as a pair.

5 Check the selector fork shaft for straightness by rolling it on a sheet of plate glass. A bent shaft will cause difficulty in selecting gears and will make the gearchange particularly heavy. Similarly, check the gearchange shaft for straightness.

6 The selector forks should be examined closely, to ensure that they are not bent or badly worn. Wear is unlikely to occur unless the gearbox

has been run for a period with a particularly low oil content. The manufacturer states that if a selector fork is only slightly bent, it is permissible to straighten it by bending it when it is cold. However, this is not recommended and renewal is advised should distortion occur. If any worn selector forks are discovered always examine the grooves in their respective pinions as well. If one component is worn it is likely that the other is also worn. If this is the case both components must be renewed as a pair.

7 The tracks in the selector drum, with which the selector forks engage, should not show any undue signs of wear unless neglect has led to under lubrication of the gearbox. If the slots in the drum are worn it must be renewed. Examine the neutral switch contact and insulating disc for damage and renew as necessary.

8 Examine the gear selector claw assembly of the gearchange shaft noting that worn or rounded ends on the claw can lead to imprecise gear selection. Similarly check the change pins in the selector drum end. The springs in the selector and detent mechanism should be unbroken and not distorted or bent in any way.

26 Gearbox shafts: dismantling and reassembly

1 The gearbox clusters should not be disturbed needlessly. They should only be stripped when careful examination of the whole assembly fails to reveal the cause of the problem, or where obvious damage such as stripped or chipped teeth is discovered.

2 The input and output shaft or main and layshaft components (as applicable) should be kept separate to avoid confusion during reassembly. Note, on 250, 251 and 300 models that the input shaft 5th gear and the output shaft 1st gear run on uncaged needle rollers, and therefore some provision must be made to catch the loose rollers as the gears are removed from the shafts. As each item is removed, place it in order on a clean surface so that the reassembly sequence is obvious and the risk of parts being fitted the wrong way round or in the wrong sequence is avoided. Care should be exercised when removing circlips to avoid straining or bending them excessively. Circlips should be opened just sufficiently to allow them to be slid off the shaft. As a safety precaution,

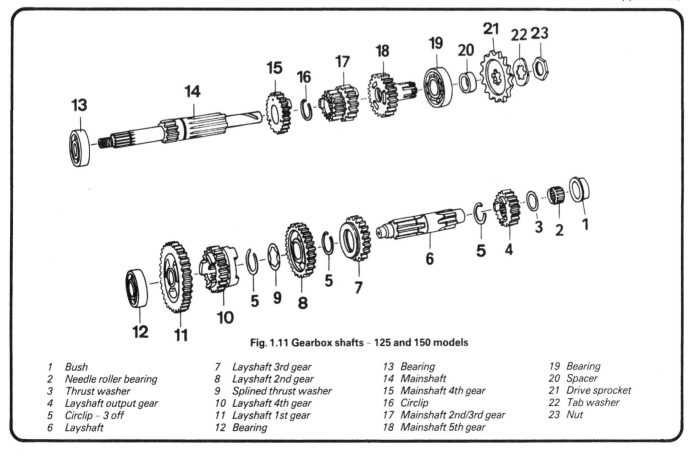

Fig. 1.11 Gearbox shafts – 125 and 150 models

1 Bush	7 Layshaft 3rd gear	13 Bearing	19 Bearing
2 Needle roller bearing	8 Layshaft 2nd gear	14 Mainshaft	20 Spacer
3 Thrust washer	9 Splined thrust washer	15 Mainshaft 4th gear	21 Drive sprocket
4 Layshaft output gear	10 Layshaft 4th gear	16 Circlip	22 Tab washer
5 Circlip – 3 off	11 Layshaft 1st gear	17 Mainshaft 2nd/3rd gear	23 Nut
6 Layshaft	12 Bearing	18 Mainshaft 5th gear	

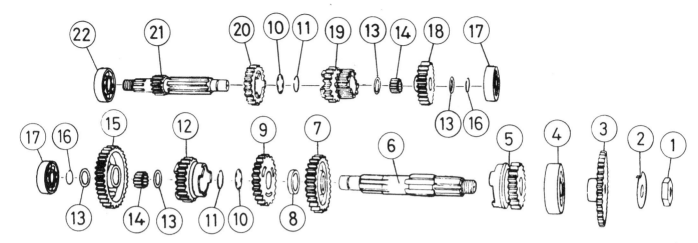

Fig. 1.12 Gearbox shafts – 250, 251 and 300 models

1   Nut
2   Tab washer
3   Drive sprocket
4   Bearing
5   Output shaft 5th gear
6   Output shaft
7   Output shaft 2nd gear
8   Spacer
9   Output shaft 3rd gear
10 Splined thrust washer – 2 off
11 Circlip – 2 off
12 Output shaft 4th gear
13 Thrust washer – 4 off
14 Needle roller – 48 off
15 Output shaft 1st gear
16 Circlip – 2 off
17 Bearing – 2 off
18 Input shaft 5th gear
19 Input shaft 2nd/3rd gear
20 Input shaft 4th gear
21 Input shaft
22 Bearing

it is recommended that all circlips are renewed regardless of their apparent condition. Examine all thrust washers for any signs of wear or damage, renewing them if necessary.

3   Having examined the gearbox components as described in the previous Section, reassemble each shaft as described below, referring to the accompanying line drawings and photographs for guidance. Note that when fitting a circlip to a splined shaft, the ends of the clip must be positioned in the middle of one of the splines to ensure that the circlip is as secure as possible on its shaft. Ensure that the bearing surfaces of each component are liberally oiled before fitting.

4   If problems arise in identifying the various gear pinions, which cannot be resolved by reference to the accompanying photographs and illustrations, the number of teeth on each pinion can be used to identify them. Count the number of teeth on the pinion and compare this figure with that given in the Specifications, remembering that the output shaft or layshaft pinions (as applicable) are listed first, followed by those on the input shaft or mainshaft. The problem of identification should not arise, however, if the instructions given in paragraph 2 of this Section are followed carefully.

Mainshaft – 125 and 150 models

5   The mainshaft is easily identified by its integral 1st gear pinion. Hold the threaded end of the shaft (left-hand end) and slide the components on from the opposite end (right-hand end) as follows.

6   Slide the 4th gear pinion along the shaft, with its dogs facing the right-hand end of the shaft, and secure it with a circlip. Slide the 2nd/3rd gear pinion onto the shaft so that the slightly smaller of the two gears

faces the 4th gear, then refit the 5th gear pinion to the right-hand end of the shaft.

Layshaft – 125 and 150 models

7   Fit a circlip to the centre of the three grooves in the shaft then hold the end of the shaft with the shorter splines (right-hand end) and slide the following components on the opposite end (left-hand end).

8   Slide the 2nd gear pinion onto the shaft, making sure its recessed face is on the left-hand side. Refit the splined thrust washer and secure the thrust washer and gear with a circlip. Slide on the 4th gear pinion, so that its selector fork groove faces the 2nd gear, then refit the 1st gear pinion ensuring that its recessed surface is on the right-hand side of the gear. Turn the shaft around and slide the following components on from the right-hand end of the shaft.

9   Fit the 3rd gear pinion, ensuring its selector fork groove faces the left-hand end of the shaft, and secure it with a circlip. Slide the output gear onto the shaft and refit the plain thrust washer. Liberally oil the needle roller bearing and fit it to the right-hand end of the shaft.

Input shaft – 250, 251 and 300 models

10   The input shaft is easily identified by its integral 1st gear pinion. Hold the threaded end of the shaft (left-hand end) and slide the components on from the opposite end (right-hand end) as follows.

11   Slide the 4th gear pinion along the shaft, with its dogs on the right-hand side of the pinion, followed by a splined thrust washer. Secure the gear and thrust washer with a circlip. Slide the 2nd/3rd gear pinion onto the shaft so that the larger of the two gears faces the 4th

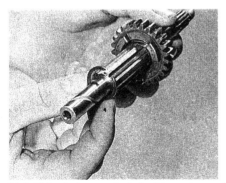

26.6a Take the bare mainshaft slide on the 4th gear pinion and secure it in position with a circlip

26.6b Refit 2nd/3rd gear pinion ...

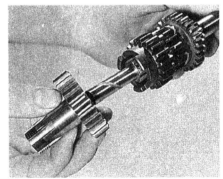

26.6c ... and fit 5th gear pinion to the end of the mainshaft

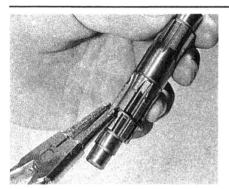

26.7 Fit a circlip to the central groove on the layshaft

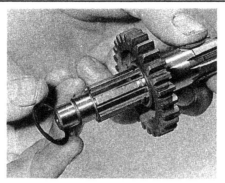

26.8a Slide 2nd gear pinion on left-hand end of the shaft followed by a splined thrust washer ...

26.8b ... and secure them both in position with a circlip

26.8c Slide on the 4th gear pinion ...

26.8d ... and the 1st gear pinion

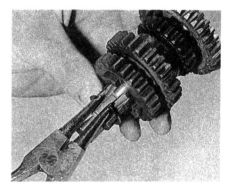

26.9a Fit the 3rd gear pinion and secure it in position with a circlip ...

26.9b Fit the output pinion to the shaft

26.9c Fit the plain thrust washer to the end of the shaft ...

26.9d ... followed by the needle roller bearing

gear pinion, followed by a plain thrust washer. Refit the 5th gear pinion so that its flat surface is on the right-hand side of the gear, then carefully insert all the needle rollers between the gear and the shaft. Slide on another plain thrust washer and secure the washer and gear in position with another circlip.

Output shaft – 250, 251 and 300 models
12   Hold the threaded end of the shaft (right-hand end) and slide the components on from the opposite end (left-hand end) as follows.
13   Slide the 2nd gear pinion onto the shaft, making sure that its recessed face is on the right-hand side. Refit the spacer and the 3rd gear pinion, ensuring that its flat surface is on the right-hand side of the gear, then slide on a splined thrust washer. Secure all the above components with a circlip. Slide the 4th gear pinion onto the shaft, so that the selector fork groove faces the 3rd gear pinion, followed by a plain thrust washer. Fit the 1st gear pinion onto the shaft, making sure that its recessed face is on the right-hand side, and carefully insert all the needle rollers into the centre of the gear. Slide on another plain thrust washer and secure

the gear and washer with a circlip. Turn the shaft around and slide the 5th gear pinion onto the right-hand end of the shaft making sure the selector fork groove is on the left-hand side of the gear.

27 Engine reassembly: general

1   Before reassembly of the engine/gearbox unit is commenced, the various component parts should be cleaned thoroughly (see Section 15) and placed on a sheet of clean paper, close to the working area.
2   Gather together all the necessary tools and have available an oil can filled with clean engine oil. Make sure that all new gaskets and oil seals are to hand, also all replacement parts required. Nothing is more frustrating than having to stop in the middle of a reassembly sequence because a vital gasket or replacement has been overlooked. As a general

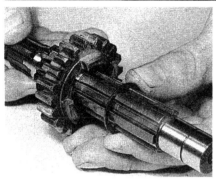

26.11a Take the bare input shaft and slide on the 4th gear pinion followed by a splined thrust washer ...

26.11b ... and secure both in position with a circlip

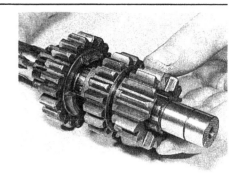

26.11c Slide on the 2nd/3rd gear pinion ...

26.11d ... and fit a plain thrust washer ...

26.11e ... and the 5th gear pinion

26.11f Refit all the needle rollers to the centre of the pinion ...

26.11g ... fit another plain thrust washer and circlip

26.13a Slide the 2nd gear pinion on the left-hand end of the output shaft followed by the spacer ...

26.13b ... 3rd gear pinion and splined thrust washer ...

26.13c ... and secure all the above components in position with a circlip

26.13d Slide on the 4th gear pinion ...

26.13e ... followed by a plain thrust washer ...

26.13f ... and the 1st gear pinion

26.13g Insert all the needle rollers into the centre of the pinion ...

26.13h ... fit another plain thrust washer and circlip

rule each moving engine component should be lubricated thoroughly as it is fitted into position.

3 Make sure that the reassembly area is clean and that there is adequate working space. Refer to the torque and clearance settings wherever they are given. Many of the smaller bolts are easily sheared if overtightened. If the existing screws show evidence of maltreatment in the past, it is advisable to renew them as a complete set.

28 Reassembling the engine/gearbox unit: preparing the crankcases

1 At this stage the crankcase castings should be clean and dry with any damage, such as worn threads, repaired. If any bearings are to be refitted, the crankcase casting must be heated first as described in Section 14.

2 Place the heated casting on a wooden surface, fully supported around the bearing housing. Position the bearing on the casting, ensuring that it is absolutely square to its housing then tap it fully into place using a hammer and tubular drift such as a socket which bears only on the outer race of the bearing. Be careful to ensure that the bearing is kept absolutely square to its housing at all times.

3 On 250, 251 and 300 models when refitting the main bearing oil guide plates ensure that the raised dot on the plate is positioned next to the oilway in the casting. Secure the plate with its circlip ensuring that the ends of the circlip are positioned on each side of the oilway. (See accompanying photo).

4 Oil seals are fitted into a cold casting in a similar manner. Apply a thin smear of grease to the seal circumference to aid the task, then tap the seal into its housing using a hammer and tubular drift which bears only on the hard outer edge of the seal, thus avoiding any risk of the seal being distorted. Tap each seal into place until its flat outer surface is just flush with the surrounding crankcase, or against its locating shoulder, as appropriate. Ensure all oil seals are fitted the same way round as those which were removed.

5 When all the bearings and seals have been fitted and secured with their retaining circlips (where fitted), lightly lubricate the bearings with

26.13i Slide the 5th gear pinion onto the right-hand end of the shaft

clean engine oil and apply a thin smear of grease to the sealing lips of each seal.

6 Support the crankcase left-hand half on two wooden blocks placed on the work surface; there must be sufficient clearance to permit the crankshaft and gearbox components to be fitted.

29 Reassembling the engine/gearbox unit: refitting the crankshaft and gearbox components

1 Refit the rotor retaining bolt to protect the end of the crankshaft and insert the crankshaft as far as possible into its main bearing in the

28.2 Tap bearings into place using a suitable tubular drift

28.3 On 250, 251 and 300 models position raised dot next to crankcase oilway

28.4 Oil seals can be drifted into position using a tubular drift

Fig. 1.13 Gear selector mechanism – 125 and 300 models

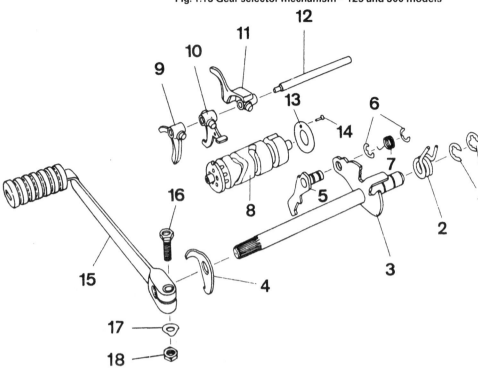

1   Circlip – 2 off
2   Return spring
3   Gearchange shaft
4   Detent arm
5   Selector claw
6   Circlip – 2 off
7   Spring
8   Selector drum
9   Selector fork – layshaft 4th
    gear
10 Selector fork – layshaft 3rd
    gear
11 Selector fork – mainshaft
    2nd/3rd gear
12 Selector fork shaft
13 Insulating disc
14 Neutral switch contact
15 Gearchange lever
16 Pinch bolt
17 Washer
18 Nut

Fig. 1.14 Gear selector mechanism – 250, 251 and 300 models

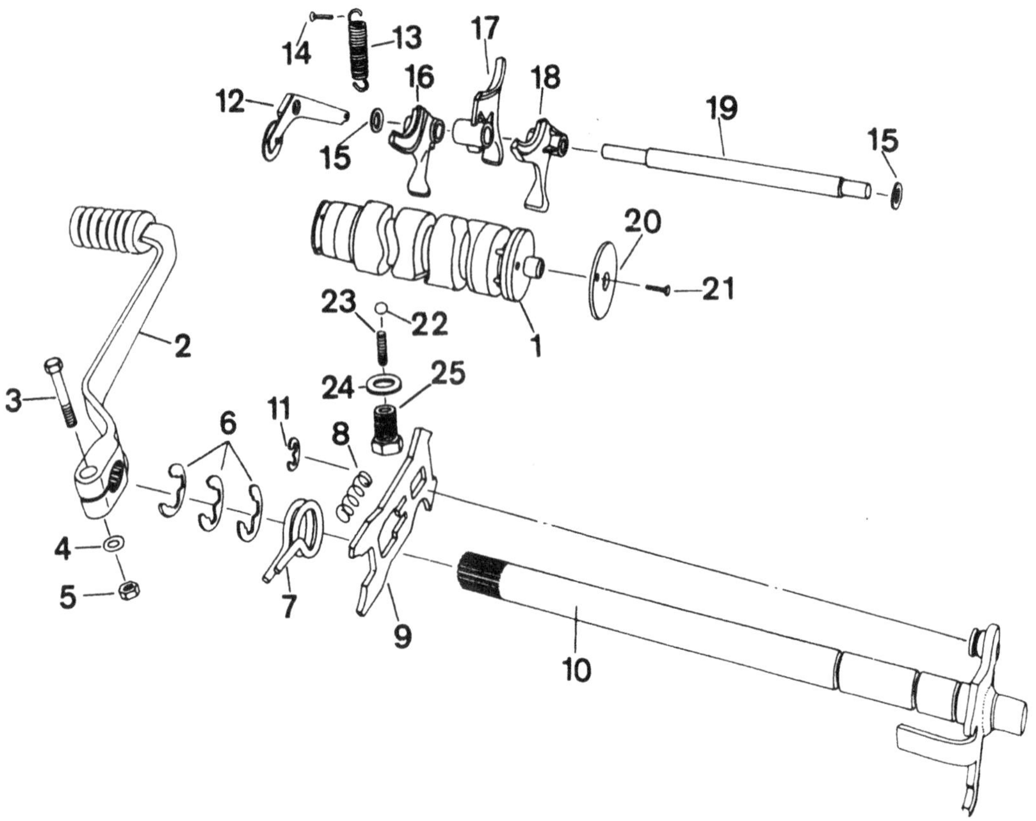

1   Selector drum
2   Gearchange lever
3   Pinch bolt
4   Washer
5   Nut
6   Circlip – 3 off
7   Return spring
8   Spring
9   Selector claw
10 Gearchange shaft
11 Circlip
12 Detent arm
13 Return spring
14 Rivet
15 Thrust washer – 2 off
16 Selector fork – output shaft
    4th gear
17 Selector fork – input shaft
    2nd/3rd gear
18 Selector fork – output shaft
    5th gear
19 Selector fork shaft
20 Insulating disc
21 Neutral switch contact
22 Neutral detent ball
23 Spring
24 Sealing washer
25 Neutral detent bolt

left-hand crankcase, using a smear of oil to ease the task. Align the connecting rod with the crankcase mouth and check that the crankshaft is square to the crankcase. Support the crank webs at a point opposite the crankpin to prevent distortion while the crankshaft is driven home with a few firm blows from a soft-faced mallet. Do not risk damaging the crankshaft by using excessive force; if undue difficulty is encountered

take the assembly to an authorized MZ dealer for the crankshaft to be pressed into place using the correct service tool.

2   Once the crankshaft is fitted, remove the protecting bolt and check that the crankshaft revolves easily with no trace of distortion.

3   Mesh both the gearbox shafts together, ensuring that all matching pinions are correctly mated, and insert the shafts into their bearings

29.1 Do not use excessive force when fitting crankshaft

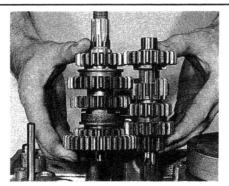

29.3 Fit gearbox shaft assemblies as a single unit

29.4 On 125 and 150 models refit all selector forks using notes made on dismantling

29.5a Refit the gearchange shaft ...

29.5b ... and install the selector drum whilst holding the gearchange selector claw

29.5c Fit selector fork pins to their respective grooves in the drum and fit the selector fork shaft

treating the two as a single assembly. If necessary, tap gently on the right-hand ends of the shafts to seat them in the casing.

125 and 150 models
4   Refit the selector forks to the grooves in their respective pinions, using the notes made on dismantling for guidance. If none were made note that on the machine shown in the photographs marks were provided in the form of a number cast on one surface of each fork. The fork marked 011 is fitted to the layshaft 4th gear pinion (bottom fork), that marked 013 is fitted to the layshaft 3rd gear pinion (top fork) and the fork marked 015 is fitted to the layshaft 2nd/3rd gear pinion (centre fork).
5   Once all the selector forks are correctly fitted insert the gearchange shaft into the casing ensuring that its return spring locates correctly with the peg in the casting. Pivot the selector forks around their respective pinions, to gain the necessary clearance to fit the selector drum, and install the selector drum whilst holding the gearchange shaft claw out of the way. Once the drum is fully in position release the claw, checking that it locates correctly with the drum, and fit the selector fork guide pins to their respective grooves in the drum. Refit the selector fork shaft.

250, 251 and 300 models
6   First position the selector fork shaft thrust washer in the casing. If necessary, the washer can be stuck to the casing with a smear of grease. Using the notes made on dismantling, refit the bottom selector fork to the output shaft 4th gear pinion and position it so that the selector drum can be installed. Once the selector drum is fully into position refit the bottom selector fork to its respective grooves in both the pinion and drum then repeat the process for both the centre and top selector forks. Once all the selector forks are in position refit the selector fork shaft, ensuring the thrust washer is still in position in the casing, and fit the second thrust washer over the end of the shaft. Refit the gearchange shaft ensuring that its return spring locates correctly with the peg in the casing and that the selector claw engages correctly with the drum.
7   On the machine featured in the photographs, the selector forks were marked with a number, cast into one surface of each fork. If no notes were taken on dismantling, or there is any confusion as to the correct position of a selector fork, these marks can be used to identify each fork. The fork marked 010 is fitted to the output shaft 4th gear

pinion (bottom fork), that marked 012 is fitted to the output shaft 5th gear (top fork) and the fork marked 011 is fitted to the input shaft 2nd/3rd gear pinion (centre fork).

All models
8   When all components are fitted, check that the selector drum is in the neutral position and that both shafts are free to rotate, then rotate the drum to check that all five gears can be selected with relative ease. Return to the neutral position and lubricate all bearings and bearing surfaces thoroughly.

---

30   Reassembling the engine/gearbox unit: joining the crankcase halves

---

1   Apply a thin film of sealing compound to the gasket surface of the left-hand crankcase half, then press the two locating dowels, if removed, firmly into their recesses in the crankcase mating surface. Apply a thin smear of sealing compound to the edge of the rubber separating disc and press it firmly into position in the left-hand crankcase half. Make a final check that all components are in position and that all bearings and bearing surfaces are lubricated.
2   Lower the upper crankcase half into position, using firm hand pressure to push it home. It may be necessary to give a few gentle taps with a soft-faced mallet to drive the casing fully into place. Do not use excessive force; instead check that all shafts and dowels are correctly fitted and accurately aligned, and that the crankcase halves are exactly square to each other. If necessary, pull away the upper crankcase half to rectify the problem before starting again.
3   When the two have joined correctly and without strain, refit the crankcase retaining screws, using the cardboard template to position each screw correctly. Working in a diagonal sequence from the centre outwards, progressively tighten the screws to the specified torque setting.
4   Wipe away any excess sealing compound from around the joint area, check the free running and operation of the crankshaft and gearbox components. If a particular shaft is stiff to rotate, a smart tap on

29.6a On 250, 251 and 300 models, position selector fork shaft thrust washer in the crankcase ...

29.6b ... and using notes made on dismantling fit bottom selector fork to the output shaft 4th gear pinion

29.6c Install the selector drum and fit selector fork guide pin to its respective groove

29.6d Fit the centre and top selector forks ...

29.6e ... and refit the selector fork shaft

29.6f Refit the gearchange shaft ensuring its return spring and claw are correctly engaged

each end using a soft-faced mallet will centralise the shaft in its bearing. If this does not work, or if any other problem is encountered, the crankcase must be separated to find and rectify the fault. If all is well, pack a clean rag into the crankcase mouth to prevent the entry of dirt, then refit the neutral detent ball bearing and spring in the bottom of the crankcase followed by the retaining bolt and washer; tighten the bolt securely.

5   On 250, 251 and 300 models refit the three rubber plugs to the right-hand side of the crankcase and refit the 8 mm bolt to the front of the crankcase, tightening it securely. On the left-hand side of the crankcase, refit the output shaft oil guide plate and cap and secure them in position with the circlip. Install the detent arm and spring, ensuring that the arm locates correctly with the selector drum, and hook the spring over its retaining pin. On 125 and 150 models refit the gearchange mechanism detent spring to the detent arm on the left-hand side of the crankcase and hook it over its retaining pin.

6   On all models, before refitting the oil seal cap(s) it is necessary to check the clearance between the outer race of the relevant bearing and

inner face of the sealing cap. This can be done, using a vernier caliper, by measuring the distance from the edge of the bearing to the sealing face of the crankcase and subtracting the height of the sealing cap ridge less the thickness of the gasket which is 0.5 mm (0.02 in). This will then give the clearance between the bearing race and sealing cap which should be 0.2 - 0.3 mm (0.008 - 0.012 in). If this is not the case the clearance can be adjusted using shims of various thicknesses which are available from any MZ dealer. Once the clearance is known to be correct, fit the required shim(s), followed by a new gasket. Apply a smear of grease to the lips of the oil seal then carefully fit the oil seal cap. Apply a sealing compound to the threads of the oil seal cap retaining screws and tighten them to their specified torque setting.

7   Once all oil seal caps have been fitted, check again that both the crankshaft and gearbox shafts are free to rotate easily. If a particular shaft is stiff to rotate, tap on smartly each end of the shaft with a soft-faced mallet to centralise the shaft on its bearings. If this does not free the shaft, remove the relevant sealing cap and recheck the clearance.

30.2 Ensure crankcase locating dowels are in position and refit right-hand crankcase half

30.5a On 250, 251 and 300 models refit output shaft oil guide plate ...

30.5b ... and cap ...

30.5c ... and secure them in position with the circlip

30.6a Refit the oil seal cap(s) using a new gasket ...

30.6b ... and tighten their retaining screws to the specified torque setting

31 Reassembling the engine/gearbox unit: refitting the tachometer drive – 125 and 150 models

1   Ensure that the circlips are correctly fitted on each end of the driven gear and refit the thrust washers to each end of the shaft. Insert the driven gear down through the casing and into position ensuring that both thrust washers remain in position on the gear.

2   Examine the O-ring which is fitted to the driven gear bush and renew it if necessary. Apply a thin smear of grease to the inside of the bush and smear the O-ring with engine oil to aid refitting. Fit the bush to the crankcase with a twisting motion as if screwing it in. When the housing is fully in place rotate it until the hole in the bush is aligned with the bolt hole in the casing. Refit the bush retaining bolt and locking tab then tighten the bolt and secure it by bending up the locking tab.

3   Slide the drive gear onto its shaft, followed by the thrust washer and secure them in position with the circlip.

32 Reassembling the engine/gearbox unit: refitting the kickstart mechanism

125 and 150 models

1   Ensure the return spring internal tang is correctly fitted in the slot on the kickstart shaft and fit the thrust washer over the gearchange shaft. Refit the kickstart shaft over the gearchange shaft and insert the external tang of the return spring into its recess in the crankcase. Temporarily fit the kickstart lever to the shaft and tension the return

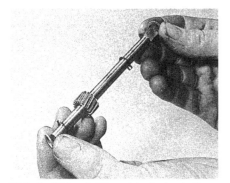

31.1a On 125 and 150 models, fit the circlips and thrust washers to each end of the tachometer driven gear ...

31.1b ... and insert gear into crankcase

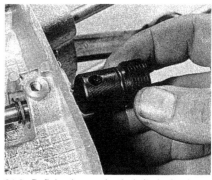

31.2a Refit bush ensuring hole in the bush aligns with bolt hole in the crankcase

31.2b Refit the bolt and secure it with the tab washer

31.3a Refit the tachometer drive gear ...

31.3b ... and secure in place with the circlip

spring by turning the kickstart shaft approximately one turn clockwise. Push the kickstart shaft fully into position, ensuring that it locates correctly with the thrust washer, and allow the spring pressure to slowly rotate the shaft anticlockwise until the quadrant is tight against its stop on the crankcase. Remove the kickstart lever.

2   The kickstart pinion is installed along with the clutch as described in Section 33.

250, 251 and 300 models

3   If the kickstart shaft was dismantled reassemble it as follows. First slide on the kickstart ratchet spring then using the marks made on dismantling fit the kickstart ratchet. If no marks were made, the pin on the ratchet must be positioned slightly to the left of the centre of the kickstart lever cotter pin flat when looking along the shaft from the right-hand (inner) end. Fit the camplate followed by a thrust washer and

the kickstart pinion. Carefully fit all the needle rollers between the pinion and shaft then slide on another thrust washer and secure all the components in position with the circlip. Fit the kickstart return spring to the opposite end of the shaft and insert the inner end of the spring into the hole in the kickstart shaft.

4   Fit the kickstart shaft into the left-hand crankcase cover, ensuring that the outer end of the return spring locates with the hole in the cover, and securely clamp the right-hand end of the kickstart shaft in a vice equipped with soft jaws. Turn the cover approximately 1¼ turns anticlockwise to tension the return spring and hold it there whilst an assistant fits the kickstart lever. Once the lever is correctly aligned with the cotter pin flat on the shaft tap the cotter pin into position, refit its nut and washer and tighten it securely. Allow the kickstart spring to slowly rotate the cover until the kickstart lever is tight against the rubber stop on the casing and remove the shaft from the vice.

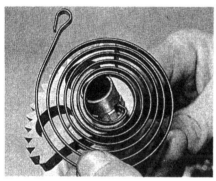

32.1a On 125 and 150 models ensure spring internal tang is correctly located in slot in kickstart shaft

32.1b Fit the thrust washer over the gearchange shaft ...

32.1c ... and refit the kickstart shaft

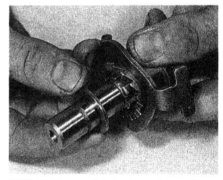

32.1d Use the kickstart lever to tension the return spring as described in text

32.3a On 250, 251, 300 models fit spring and ratchet to kickstart shaft – use marks made on dismantling

32.3b Fit the camplate to the ratchet and slide on a plain thrust washer ...

32.3c ... and the pinion

32.3d Insert all the needle rollers into the pinion ...

32.3e ... and fit another plain thrust washer and a circlip

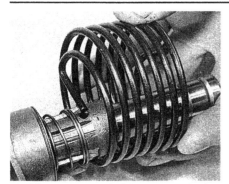

32.3f Ensure inner end of the return spring is fitted into the hole in the kickstart shaft

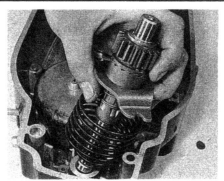

32.4a Ensure outer end of the return spring fits into the hole in the cover as the shaft is fitted

32.4b Tension return spring as described in text, refit cotter pin and tighten its retaining nut securely

33   Reassembling the engine/gearbox unit: refitting the clutch and primary drive components

125 and 150 models

1   Before fitting the primary drive components it is necessary to check that the primary drive and driven sprockets are perfectly aligned. To do this, fit the clutch drum thrust washer, shims (where fitted) and collar to the mainshaft and slide on the clutch drum. Fit the primary drive sprocket to the end of the crankshaft. Ensure both the drum and sprocket are pushed fully onto their shafts then check that both are correctly aligned by means of a straightedge (see accompanying photo). If not correctly aligned, the primary drive chain and sprockets will wear prematurely. If necessary, adjustment can be made by the addition or subtraction of shims fitted between the clutch drum thrust washer and collar. These shims can be purchased from any MZ dealer and are available in the following thicknesses: 0.1, 0.2, 0.3 and 0.5 mm.

2   Once both components are correctly aligned remove both the clutch drum and primary drive sprocket from their shafts, fit them both to the primary drive chain and install all components as an assembly. Refit the primary drive sprocket bolt along with its washers. If the top end of the engine has been removed, lock up the crankshaft by passing a close-fitting round bar through the connecting rod small-end eye and resting the bar on two wooden blocks placed across the crankcase mouth; tighten the primary drive sprocket bolt to the specified torque setting. Alternatively, if the engine is in the frame the crankshaft can be locked using the method employed on dismantling once the clutch has been assembled (see Section 9).

3   Slide a thrust washer along the mainshaft followed by the clutch centre, a new tab washer and the clutch centre nut; tighten down the nut as far as possible by hand, noting that it has a left-hand thread and must be tightened anticlockwise.

4   Lock the clutch centre using the method employed on dismantling and tighten the retaining nut to the specified torque setting. Secure the nut by bending the tab of the washer against one of the flats of the nut. Check that the clutch centre and outer drum can rotate freely, easily and independently of one another.

5   The clutch plain and friction plates should be coated with transmission oil prior to installation. Refit the thick, tapered plain plate first then install the friction and plain plates alternately until all are in position, then refit the mushroom-headed pushrod. Refit the pressure plate ensuring that the raised dot cast into the plate aligns with the line on the clutch centre (see accompanying photo). Fit the clutch spring and support plate, noting that the spring must be fitted with its convex side facing outwards (away from the crankcase) .

6   Before fitting the clutch bolts it is necessary to measure the distance from the edge of the clutch centre to the outer surface of the support plate. This distance is then used to calculate the thickness of triangular shims required to ensure the smooth operation of the clutch. This measurement must be taken when all the clutch components are pressed lightly together, removing all freeplay, but without compressing the clutch spring. Subtract the measurement taken from 20.7 mm (0.81 in), the resulting figure being the correct thickness of shims required. These shims can be purchased from any MZ dealer and are available in the following thicknesses: 0.1, 0.2, 0.3, 0.5 and 1.0 mm.

7   Position the necessary shims on the support plate, along with three new tab washers, and refit the clutch bolts. Ensure the tang on each tab washer is hooked over the inside of the triangular shim(s) and then tighten the clutch bolts evenly and progressively to their recommended torque setting. Secure each bolt by bending up a portion of the tab washer against one of its flats .

250, 251 and 300 models

8   If the clutch was dismantled it must be reassembled as follows. Liberally coat the clutch plain and friction plates in transmission oil then fit them alternately into the clutch body, starting and finishing with a friction plate. Align all the friction plate grooves, insert the small primary drive gear and refit the pressure plate, ensuring the marks made on dismantling are correctly aligned. Turn the clutch body over and fit the six plain washers over the pressure plate studs and position the clutch springs in the circular cut outs in the body. Refit the pressure flange, using the marks made on dismantling, and compress the assembly in a vice equipped with soft jaws. Fit new tab washers over the pressure plate studs and refit the six clutch nuts. Tighten the nuts evenly and

33.1a Refit the thrust washer, shims (where fitted) ...

33.1b ... and collar to the mainshaft

33.1c Fit clutch drum and drive sprocket and check that both sprockets are correctly aligned

33.2a Fit clutch drum, drive sprocket and chain as an assembly

33.2b Refit the drive sprocket retaining bolt and washers ...

33.2c ... and tighten it to specified torque setting whilst locking crankshaft as described in text

33.3a Fit the thrust washer ...

33.3b ... and clutch centre. Fit a a new tab washer ...

33.4a ... and tighten centre nut to specified torque setting whilst holding clutch centre as shown

33.4b Bend tab washer up against one of the flats on centre nut

33.5a Fit thick plain plate first, ensuring its tapered surface is facing the crankcase ...

33.5b ... then install the friction ...

33.5c ... and plain plates alternately ...

33.5d ... and refit the mushroom-headed pushrod

33.5e Refit the pressure plate ...

33.5f ... ensuring that raised dot aligns with the line on clutch centre

33.5g Fit clutch spring with its convex side facing outwards ...

33.5h ... and the support plate

33.7a Calculate and fit the required amount of shims ...

33.7b ... refit the clutch bolts and tab washers, tightening them to the specified torque setting ...

33.7c ... and bend up the tab washers

progressively until all the nuts are securely tightened then remove the assembly from the vice. Secure each nut in position by bending up the tab washer against one of its flats and remove the primary drive gear from the clutch assembly.

9   Refit the large primary driven gear to the input shaft splines and tap it fully onto the shaft using a soft-faced mallet. Fit a new tab washer, ensuring that its locating tang is correctly inserted in the cutout of the gear, and tighten the gear retaining nut down as far as possible by hand. Prevent the gear from rotating, using the method employed on dismantling, and tighten the retaining nut to its specified torque setting. Secure the nut in position by bending up a portion of the tab washer against one of its flats.

10   Before fitting the primary drive gear to the crankshaft, note that both thrust washers fitted each side of the gear have a tapered inner circumference. Both washers must be fitted with the taper facing the crankcase, so that the taper is against the steps on the crankshaft. Slide on the larger thrust washer followed by the needle roller bearing.

Liberally oil the needle roller bearing and refit the primary drive gear followed by the smaller thrust washer and spring washer.

11   Before proceeding further it is necessary to check the endfloat of the primary drive gear. To do this accurately will require the use of the special MZ service tool and a DTI gauge (Dial Test Indicator). If this equipment is not available it will be necessary to have the endfloat checked by an MZ dealer. Alternatively a spacer could be fabricated that will clamp the small thrust washer firmly when the nut is fitted to the end of the crankshaft, therefore holding the gear in place. Then position a DTI gauge so that it contacts the top surface of the primary gear and measure the endfloat of the gear. The endfloat should be between 0.05 – 0.10 mm (0.002 – 0.004 in) and is adjusted by changing the large thrust washer fitted behind the primary drive gear. This washer is available in the following thicknesses: 1.90, 1.95 and 2.00 mm.

12   Align the clutch assembly with the primary drive gear and push the clutch onto the crankshaft taper. Before fitting the left-hand crankcase cover it is necessary to press the clutch assembly fully onto the

33.8a On 250, 251 and 300 models alternately fit the plain ...

33.8b ... and friction plates to the clutch body

33.8c Fit the primary drive gear to align all friction plate slots ...

33.8d ... and refit the pressure plate

33.8e Fit the six washers and clutch springs

33.8f Refit the pressure flange

33.8g Tighten all clutch nuts securely and bend up tab washers

33.9a Refit large primary driven gear and fit a new tab washer

33.9b Tighten nut to specified torque setting whilst retaining gear as shown ...

33.9c ... and secure it with the tab washer

33.10a Slide the larger thrust washer along the crankshaft followed by the needle roller bearing

33.10b Refit the primary drive gear and the smaller thrust washer and spring washer

33.12a Refit the clutch assembly

33.12b Press clutch onto crankshaft taper as described in text

33.13a Fit the external worm gear to the casing ...

33.13b ... and refit the internal worm gear

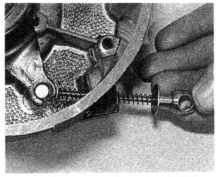
33.13c Refit the connecting rod to the lifting arm ...

33.13d ... and adjust lifting mechanism as described in text

crankshaft. To do this a spacer is needed which is exactly the same diameter and width as the inner race of the bearing fitted to the clutch release mechanism, ideally the bearing itself could be used if it has been removed from the casing. Fit the spacer over the end of the crankshaft, refit the nut to the left-hand end of the crank and tighten it as far as possible by hand. Lock up the crankshaft, tighten the nut to the specified torque setting, then slacken and remove the nut and spacer.

13 If the clutch release mechanism was removed for examination it must be refitted as follows. Press the external worm gear and bearing of the release mechanism fully into place in the left-hand crankcase cover. If necessary, tap the gear into position using a hammer and suitable drift. Engage the internal worm gear with the external gear and refit the connecting rod to the lifting arm. Adjust the clutch lifting mechanism so that when the worm gear is fully home there is approximately 6 mm (0.24 in) of clearance between the lifting arm and the lug on the inside of the crankcase cover (see accompanying photo).

34 Reassembling the engine/gearbox unit: refitting the left-hand crankcase cover

1 Make a final check that all components are correctly in position, securely fastened and lubricated. Ensure the two locating dowels are in place in the gasket face of the crankcase left-hand half. Check the condition of the O-ring in the crankcase cover, renewing it if necessary. On 125 and 150 models refit the thrust washer to the kickstart shaft.

2 Thoroughly clean both gasket surfaces and place a new gasket on the crankcase sealing face, using a smear of grease to stick it in place. Smear a small amount of grease on the crankcase cover O-ring and liberally grease the splines of the kickstart or gearchange shaft (as necessary). Offer up the cover, taking great care not to damage the O-ring as the shaft splines pass through it. On 250, 251 and 300 models it is necessary to align the kickstart camplate with its recess in the crankcase as the cover is fitted. This task can be simplified by using the

lower rear cover retaining screw to hold the plate in position as the cover is fitted .

3 On all models, once the cover is pushed fully home, refit the retaining screws in their correct locations, not forgetting the oil feed pipe guide (where fitted) on 125 and 150 models or the sealing washer which is fitted to the cover drain screw on 250, 251 and 300 models. Tighten the cover screws evenly and progressively up to their recommended torque setting. On 250, 251 and 300 models refit the tacho-meter drive gear or nut (as appropriate) and flat washer to the end of the crankshaft and tighten it to its specified torque setting whilst preventing crankshaft rotation using the method employed on dismantling.

4 If the work is being carried out with the engine/gearbox unit in the frame, it will now be necessary to refit the remaining components listed in Section 8, the necessary work being described in Sections 37 and 38.

35 Reassembling the engine/gearbox unit: refitting the generator

1 Degrease the rotor and crankshaft tapers and refit the Woodruff key or pin (as applicable) into the crankshaft. Align the slot in the rotor with the Woodruff key or pin in the crankshaft and push the rotor onto the crankshaft. Gently tap the rotor centre with a soft-faced mallet to seat it then fit the contact breaker cam and washer, ensuring that its slot locates correctly with the tab on the rotor, and the spring washer and bolt. Tighten the rotor retaining bolt to the specified torque setting.

2 Align the slot in the stator housing with the pin in the crankcase and refit the stator housing. Refit the three stator housing retaining screws and tighten them to their specified torque setting.

3 Refit the carbon brushholder to the stator housing, refit the holder retaining screws, and tighten them securely. Connect the brush wires to the relevant terminals of the stator housing.

4 If the work is being carried out with the engine/gearbox unit in the frame, reconnect the wiring to the stator housing terminals, using the notes made on dismantling, and refit the right-hand crankcase cover. Refer to Section 37 of this Chapter for further information.

34.1a Check condition of cover O-ring, renewing if necessary

34.1b On 125 and 150 models do not forget to refit thrust washer on kickstart shaft

34.2a On 250, 251 and 300 models use screw (arrowed) to ensure ...

34.2b ... kickstart camplate fits into its recess in crankcase as cover is fitted

34.3a On 125 and 150 models do not omit oil pipe guide (where fitted)

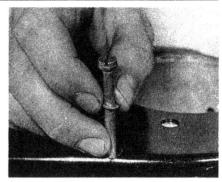

34.3b On 250, 251 and 300 models do not omit sealing washer from cover drain screw

35.1a Fit the Woodruff key or pin (as applicable) ...

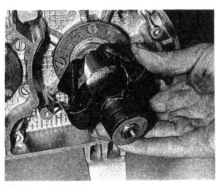

35.1b ... and refit the rotor

35.1c Fit the cam and rotor retaining bolt ...

35.1d ... ensuring slot in cam locates with tab on the rotor

35.2a Align slot in stator housing with pin in crankcase when fitting stator

35.2b Tighten stator housing screws to specified torque setting

36 Reassembling the engine/gearbox unit: refitting the piston, cylinder barrel and head

1   The piston rings are refitted to the piston using the same technique as on removal. If the original rings are to be re-used, check that they are fitted the same way up as before and in their original grooves. Ensure that the ring end gaps are located correctly by their retaining pegs in the piston.

2   Remove the rag used to plug the crankcase mouth. Lubricate the crankshaft main bearings and big-end bearing with fresh two-stroke oil, then lubricate the small-end bearing and insert it into the connecting rod small-end eye. Check that the sealing surfaces of the crankcase mouth and the cylinder barrel are perfectly clean and free from grease, oil or dirt. Check that the cylinder mounting studs are securely tightened into the crankcase. Temporarily replace the rag in the crankcase mouth whilst the gudgeon pin circlips are fitted.

3   Position the piston over the connecting rod small-end eye so that the arrow cast in the piston crown points to the front, or towards the exhaust port, and secure it by pushing the gudgeon pin through the small-end bearing and the piston boss. Fit new circlips ensuring that they are correctly seated.

4   Check that the piston rings are still correctly positioned in their piston grooves and then lubricate the piston rings and cylinder bore with fresh two-stroke oil. Fit a new base gasket and lower the cylinder barrel down over the studs and onto the piston whilst pressing the ring ends together. This should not require excessive force so if any difficulty is encountered remove the barrel and double check that the rings are correctly fitted before trying again. Once the rings are fully engaged in the bore, remove the rag from the crankcase mouth and push the barrel carefully down to rest on the crankcase mouth.

5   Rotate the crankshaft so that the piston is at the top of its stroke and wipe away any surplus oil. Check that the mating surfaces of the cylinder head and barrel are clean and dry and place a new cylinder head gasket of the required thickness over the studs. Refit the cylinder head, washers and nuts. Tighten the cylinder head nuts evenly and in a

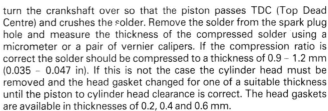

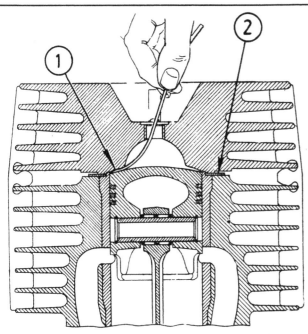

Fig. 1.15 Measuring the piston/cylinder head clearance (Sec 36)

1   *Lead solder*   2   *Cylinder head gasket*

diagonal sequence, working up in stages until the specified torque setting is reached .

6   Once the cylinder head nuts have been torqued down it is necessary to check the compression ratio of the engine. This is done using a piece of lead solder which has a diameter of approximately 2 mm (0.08 in) to measure the piston to cylinder head clearance. Bend the solder into a slight curve and insert it into the spark plug hole in the cylinder head until the end of the solder is touching the cylinder wall. Sharply

turn the crankshaft over so that the piston passes TDC (Top Dead Centre) and crushes the solder. Remove the solder from the spark plug hole and measure the thickness of the compressed solder using a micrometer or a pair of vernier calipers. If the compression ratio is correct the solder should be compressed to a thickness of 0.9 – 1.2 mm (0.035 – 0.047 in). If this is not the case the cylinder head must be removed and the head gasket changed for one of a suitable thickness until the piston to cylinder head clearance is correct. The head gaskets are available in thicknesses of 0.2, 0.4 and 0.6 mm.

7   Once the compression ratio of the engine is known to be correct, refit the heat dissipator (where fitted) to the cylinder head and refit the four rubber noise dampers to the cooling fins of the head and barrel. Clean the spark plug and check its gap as described in Routine maintenance. Smear graphited grease over its threads and screw it into the cylinder head.

8   If the work is being carried out with the engine/gearbox unit in the frame it will now be necessary to reassemble all those components listed in Section 6, the necessary work being described in Sections 37 and 38.

37 Refitting the engine/gearbox unit to the frame

1   The engine/gearbox unit is refitted to the frame by reversing the removal sequence. Ensure the machine is securely positioned on its centrestand before starting work and apply a smear of grease to the mounting bolts. With the aid of an assistant, offer up the engine and refit both the rear mounting bolts, followed by the nuts and spring washers which secure the cylinder head to its rubber mounting. On 250, 251 and 300 models tighten the engine mounting nuts and bolts to the specified torque setting. On 125 and 150 models tighten all nuts and bolts securely, then refit the rubber plugs which obscure the rear engine mounting bolts.

2   Grease the lips of the mainshaft or output shaft (as applicable) oil seal and on 125 and 150 models carefully slide the spacer along the mainshaft and insert it into the seal. On all models, engage the sprocket on the chain and fit the sprocket over the shaft splines. Refit the tab

36.1 Ensure piston ring end gaps align with piston locating pegs

36.2 Lubricate all bearings thoroughly and pack crankcase mouth with clean rag

36.3a Refit piston ...

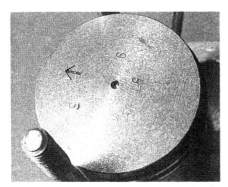

36.3b ... ensuring arrow on piston crown faces forward ...

36.3c ... and fit new circlips

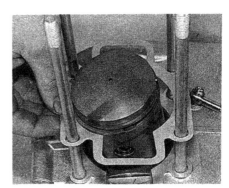

36.4a Fit a new base gasket ...

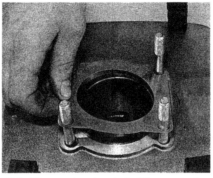

36.4b ... and lower the cylinder barrel over the piston

36.5a Fit a new head gasket of the required thickness ...

36.5b ... then refit cylinder head and tighten nuts to specified torque setting

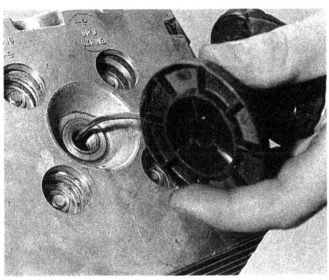

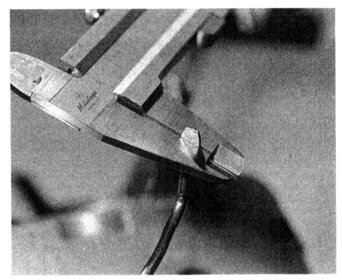

36.6a Check piston cylinder head clearance as described in text ...

36.6b ... to ensure compression ratio is correct

washer and tighten the sprocket retaining nut to the specified torque setting whilst applying the rear brake hard to prevent the sprocket from rotating (engine in gear). Secure the nut by bending the locking tab up against one of the flats on the nut. Check the final drive chain tension, rear brake adjustment and stoplamp switch setting as described in Routine maintenance and adjust as necessary. On 125 and 150 models smear a small amount of grease to the inner end of the clutch pushrod and insert it into the end of the mainshaft .

3   Feed the alternator wiring through the hole in the crankcase and, using the notes made on dismantling, refit the connectors to their relevant terminals on the stator housing. Reconnect the wire to the neutral switch (where fitted). Ensure that the wiring is correctly routed and is in no danger of being trapped or coming in contact with the chain and that the wiring loom sealing grommet is correctly fitted to the crankcase.

4   On 250, 251 and 300 models slide the clutch mounting sleeve along the clutch cable and reconnect the cable to the clutch lifting mechanism connecting rod. Refit the clutch adjusting plate and O-ring and adjust the clutch lifting mechanism as described in Routine maintenance. Refit the tachometer drive housing or cover (as applicable), tightening its retaining screws securely, and complete clutch adjustment by setting the cable freeplay.

5   Connect the tachometer cable (where fitted) to its drive and tighten its retaining ring securely. Refit the carburettor, exhaust and oil pump (where fitted) as described in the relevant Sections of Chapter 2. Working as described in Routine maintenance, check the throttle cable and oil pump cable adjustment, and remove any air from the oil pump and feed line using the bleed screw as described in Chapter 2. Using the marks made on dismantling, refit the gearchange and kickstart lever (as appropriate) and tighten its/their pinch bolt(s) securely.

6   Check and, if necessary, adjust the ignition timing as described in Routine maintenance. Refit the right-hand crankcase cover, ensuring that the chain gaiters are correctly positioned, and tighten its retaining screws to the recommended torque setting. On 125 and 150 models, once the cover has been fitted, check and adjust the clutch cable as described in Routine maintenance.

7   Connect the battery to the main loom (negative terminal last), ensuring that its terminals are free from corrosion. Check that the transmission oil drain plugs are securely fastened and, referring to Routine maintenance, pour in the correct amount of the specified oil; if this amount cannot be measured accurately use the level screw as described in Routine maintenance. Note that the level will drop after the engine has been run for the first time as the oil is distributed around the various components and will therefore require careful rechecking. Refit the filler plug.

8   Make a final check that all components have been refitted and that all are correctly adjusted and securely fastened; the exception being the oil pump cover (where fitted).

38   Starting and running the rebuilt engine

1   Attempt to start the engine using the usual procedure adopted for a cold engine. Do not be disillusioned if there is no sign of life initially. A certain amount of perseverance may prove necessary to coax the engine into activity even if new parts have not been fitted. Should the engine persist in not starting, check that the spark plug has not become fouled by the oil used during reassembly. Failing this go through the

fault diagnosis section and work out what the fault is methodically.

2   When the engine does start, keep it running as slowly as possible to allow the oil to circulate. Open the choke as soon as the engine will run without it. During the initial running, a certain amount of smoke may be in evidence due to the oil used in the reassembly sequence being burnt away. The resulting smoke should gradually subside.

3   Check the engine for blowing gaskets and oil leaks. Before using the machine on the road, check that all the gears select properly, and that the controls function correctly.

4   As soon as the engine reaches normal operating temperature adjust the carburettor settings as described in Section 8 of Chapter 2. Once the carburettor settings have been adjusted and the machine is idling smoothly, stop the engine and check the transmission oil level as described in Routine maintenance. If necessary, add oil until the level is correct then refit the level screw, tightening it securely, and filler plug. Finally recheck the oil pump operating cable setting as described in Routine maintenance, adjusting it as necessary, then refit the oil pump cover and tighten its retaining screws securely.

39   Taking the rebuilt machine on the road

1   Any rebuilt machine will need time to settle down, even if parts have been refitted in their original order. For this reason it is highly advisable to treat the machine gently for the first few miles to ensure oil has circulated throughout the lubrication system and that new parts fitted have begun to bed down.

2   Even greater care is necessary if the engine has been rebored or if a new crankshaft has been fitted. In the case of a rebore, the engine will have to be run in again, as if the machine were new. This means greater use of the gearbox and a restraining hand on the throttle until at least 500 miles have been covered. There is no point in keeping to any set speed limit; the main requirement is to keep a light loading on the engine and to gradually work up performance until the 500 mile mark is reached. These recommendations can be lessened to an extent when only a new crankshaft is fitted. Experience is the best guide since it is easy to tell when an engine is running freely.

3   Remember that a good seal between the piston and cylinder bore is essential for the correct functioning of the engine. A rebored two-stroke will require more careful running in, over a longer period, than its four-stroke counterpart. There is a far greater risk of engine seizure during the first hundred miles if the engine is permitted to work hard.

4   If at any time a lubrication failure is suspected, stop the engine immediately and investigate the cause. If an engine is run without oil, even for a short period, irreparable engine damage is inevitable.

5   On models with oil-injection, do not on any account add oil to the petrol under the mistaken belief that a little extra oil will improve the engine lubrication. Apart from creating excess smoke, the addition of oil will make the mixture much weaker, with the consequent risk of overheating and engine seizure. The oil pump alone should provide full engine lubrication.

6   Do not tamper with the exhaust system. Unwarranted changes in the exhaust system will have a marked effect on engine performance, invariably for the worse. The same advice applies to dispensing with the air filter or its element.

7   When the initial run has been completed allow the engine to cool down and then check all the fittings and fasteners for security. Re-adjust any controls which may have settled down during initial use and re-check the transmission oil level.

37.2a Tighten sprocket retaining nut to specified torque setting and secure with the tab washer

37.2b On 125 and 150 models insert the pushrod into mainshaft

37.3 Ensure generator wiring is correctly routed and in no danger of being trapped or chafed

37.5a Refit tachometer cable (where fitted) – 125 shown

37.5b Using the marks made on dismantling refit the kickstart and gearchange levers (as applicable) – 125 shown

37.6 Check ignition timing and refit right-hand crankcase cover

Chapter 2 Fuel system and lubrication

Refer to Chapter 7 for information relating to 1991-on models

Contents

Specifications

Fuel tank capacity
Total including reserve:
- 125 and 150 models .... 13 lit (2.8 Imp gal)
- 250, 251 and 300 models .... 17 lit (3.7 Imp gal)
- Reserve – all models .... 1.5 lit (0.3 Imp gal)

Recommended fuel grade
.... Unleaded or low-lead, minimum octane rating 88 RON/RM

Carburettor

	ETZ125 models	ETZ150 models
Type	BVF 22N 2-2	BVF 24N 2-2
Main jet	100	120
Needle jet	2.5 A 513	2.5 A 513
Needle clip position – grooves from top*	3rd – 5th	3rd – 5th
Starter jet	70	
Pilot jet	35	40
Pilot screw – turns out	$1\frac{1}{2}$	$1\frac{1}{2}$

Carburettor

	ETZ250 (early) models	ETZ250 (late), 251 and 300 models
Type	BVF 30 N 2-5	BVF 30 N 3-1
Main jet	130 (125 on some early models)	130
Needle jet	70	70
Jet needle	2.5 A 512	2.5 B 511
Needle clip position – grooves from top*	3rd – 5th	3rd – 5th
Starter jet	110	95
Pilot jet	45	50
Pilot screw – turns out	$2 – 2\frac{1}{2}$	$2\frac{1}{2}$
By-pass air screw	Not available	4

* 5th groove from the top of needle jet for running in

Engine lubrication – pre-mix models
- Recommended oil .... Good quality self-mixing two-stroke oil
- Fuel/oil pre-mix ratio .... 50:1 (0.16 pint/136 cc oil to 1 gal petrol, 30 cc oil to 1 litre petrol)

Engine lubrication – oil-injection models
- Recommended oil .... Good quality branded two-stroke oil
- Oil tank capacity .... Approximately 1.3 litres (0.27 Imp gal)

Transmission lubrication
Capacity:
- 125 and 150 models .... 500 cc (0.88 Imp pint)
- 250, 251 and 300 models .... 900 cc (1.58 Imp pint)
- Recommended oil – all models .... Good quality EP80 hypoid gear oil

Torque settings
- Exhaust pipe ring nut .... 15.0 kgf m (108.0 lbf ft)

1 General description

Fuel from the frame-mounted tank is fed to the float chamber of the BVF carburettor via a three position fuel tap. Air is drawn into the carburettor through an air filter housing which contains a pleated dry paper type element. The proportions of air and atomized fuel are regulated by the carburettor, which provides the correct mixture at all throttle positions. Cold starting is assisted by a manual choke which is cable operated from a handlebar-mounted lever.

Engine lubrication varies depending on the model. Some models use a 'petroil' system in which the oil and petrol are mixed in a specified ratio whenever the tank is refilled. The oil is retained in the mixture as it passes through the carburettor and lubricates the various components as it circulates around the engine. Other models employ oil-injection, where two-stroke oil is gravity fed from a frame mounted tank to the oil pump mounted on the left-hand side of the crankcase. A cable connecting the oil pump and throttle twistgrip ensures the delivery of the correct quantity of oil as the engine speed varies. The oil is then pumped directly into the engine via the crankcase.

Gearbox and primary drive components are lubricated by a splash feed from a supply of oil in the reservoir formed by the crankcase castings.

The exhaust system is a two-part assembly comprising a downpipe and silencer.

2 Fuel tank: removal, examination and refitting

Note: *Petrol is extremely flammable, particularly in the form of vapour. Take all precautions to prevent the risk of fire and read the Safety first! section of this manual before starting work.*

1 Slacken and remove the two nuts and washers which secure the rear of the seat to the frame and one retaining strap bolts, and lift the seat away from the machine.

2 Turn the fuel tap to the OFF position and disconnect the fuel pipe from the tap. Slacken and remove the nut and bolt securing the rear of the fuel tank to the frame, and remove the washer and collar from the rubber mounting. Remove both the rear mounting rubbers, noting how they are positioned, then pull the tank rearwards until it is freed from its front mounting rubbers, and lift it clear of the machine.

3 Store the tank upright in a safe place whilst removed from the machine, well away from any naked flames or lights. Check that the tap is not leaking and that it cannot be accidentally knocked into the ON position. Covering the tank with a soft protective cloth may well avoid damage being caused to the finish by dirt, grit, dropped tools, etc.

4 If fuel tank repairs are necessary the following points should be noted. Fuel tank repair, whether necessitated by accident damage or fuel leakage, is a task for the professional. Welding or brazing is not

recommended unless the tank is first purged of all fuel vapour; which is not an easy condition to achieve. Resin-based tank sealing compounds are a much more satisfactory method of curing leaks, and are available at most auto accessory shops. Accident damage repairs will inevitably involve re-painting the tank. Matching of modern paint finishes, especially metallic ones, is a very difficult task and is not to be lightly undertaken by the average owner. It is therefore recommended that the tank be removed by the owner, and taken to a motorcycle dealer or similar expert for expert professional attention.

5 Repeated contamination of the fuel tap filter and carburettor by water or rust and paint flakes indicates that the tank should be removed for flushing with clean fuel and internal inspection. Rust problems can be cured by using a resin-based tank sealant.

6 On refitting, examine all rubber tank mountings and renew any which show signs of compression or deterioration. Apply a small amount of grease to the front mounting rubbers and refit the tank, taking care not to trap the control cables. Ensure the tank is correctly located on the front mounting rubbers and refit the rear mounting rubbers to their original positions. Insert the collar and refit the mounting bolt, flat washer and nut, tightening them securely. Refit the seat and tighten both its retaining nuts and washers and the bolt which secures the retaining strap to the frame. Finally reconnect the fuel pipe to the tank and turn the tap on. Thoroughly check for leaks before taking the machine on the road.

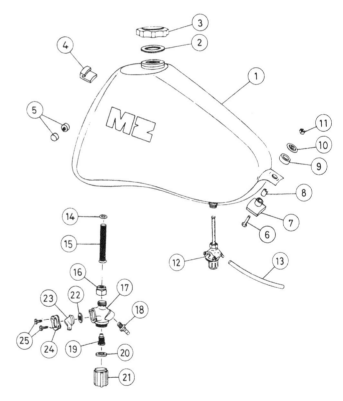

Fig. 2.1 Fuel tank and tap

1 Fuel tank	14 Gasket
2 Gasket	15 Fuel filter
3 Filler cap	16 Mounting nut
4 Damping rubber	17 Tap body
5 Mounting rubber – 2 off	18 Pipe spigot
6 Bolt	19 Fuel filter
7 Damping rubber	20 Gasket
8 Spacer	21 Filter bowl
9 Damping rubber	22 Tap seal
10 Washer	23 Tap lever
11 Nut	24 Retaining plate
12 Fuel tap	25 Screw – 2 off
13 Fuel pipe	

2.6 Do not omit mounting rubbers when refitting fuel tank

3   Fuel tap: removal, examination and refitting

Note: *Petrol is extremely flammable, particularly in the form of vapour. Take all precautions to prevent the risk of fire and read the Safety first! section of this manual before starting work.*

1   To remove the fuel tap, first drain the contents of the tank into a clean container which is suitable for carrying petrol. This can be done by disconnecting the fuel pipe from the tap, placing the container underneath the tap, and turning the tap to the RES position. Once the fuel has drained from the tank, unscrew its retaining nut and remove the tap from the machine.

2   Unscrew the mounting nut and remove the tap sealing washer from the tap body. Remove the upper fuel filter and nylon stack pipe, noting which hole the pipe is fitted to. Unscrew the filter bowl from the bottom of the tap and remove its sealing washer. The lower fuel filter can then be unscrewed. Release the two screws which retain the tap lever retaining plate, withdraw the plate, lever and tap seal. The fuel pipe spigot can also be unscrewed from the body of the tap if required.

3   Thoroughly clean all components and check for signs of wear or damage, especially the mating surfaces of the tap body and lever, and the seal itself. If the tap lever retaining screws have been overtightened, the seal will be distorted and must be renewed to prevent fuel leakage or restriction in the fuel supply; this would cause a weak mixture, due to a lack of fuel in the carburettor, and would lead to a drop in performance or even engine damage. MZ state that the flow rate through the fuel tap must be at least 12 litres (2.64 gal) per hour. Wash both the fuel filters in clean petrol and examine them for cracks and splits. If either filter is damaged or badly blocked it must be renewed. Examine the condition of the tap sealing washers and renew them if necessary. Note that if bolt filters show a heavy build-up of rust or corrosion, the fuel tank should be thoroughly cleaned out as described in the previous section.

4   On reassembly, fit the tap seal, aligning it carefully with the passages in the tap body. Refit the lever and retaining plate. *Do not overtighten its retaining screws; these should be tightened sufficiently to hold the lever against the seal and prevent fuel leakage.* Note that there should be a distinct gap between the retaining plate and tap body. Refit the lower fuel filter and sealing washer and tighten the filter bowl securely. Refit the nylon stack pipe to its original position in the tap body and install the upper fuel filter. Fit the sealing washer to the tap mounting nut and refit the nut to the tap noting that its hexagonal end must face the tap body, and refit the pipe spigot.

5   Refit the fuel tap to the tank and tighten its mounting nut just enough to compress the sealing washer. Connect the fuel pipe to the tap spigot and refill the tank with fuel. Turn the fuel supply on and thoroughly check for leaks before taking the machine out on the road.

4   Fuel pipe: examination

Note: *Petrol is extremely flammable, particularly in the form of vapour. Take all precautions to prevent the risk of fire and read the Safety first! section of this manual before starting work.*

1   The fuel pipe is constructed of a tough but flexible petrol-proof

material and has an internal bore of 5 mm (0.20 in). The seal at each end of the pipe being effected by the interference fit between the pipe and spigot.

2   Check the pipe regularly looking for signs of splits, cracks or chafing. If the pipe is only damaged at the ends a temporary repair can be effected by cutting off the damaged portion and refitting the pipe. Obviously this can be done only if there is a sufficient length of pipe.

3   If renewal is necessary use only the original type of tubing, available from MZ dealers. When renewing the pipe it is advisable to purchase a length of pipe which is one or two inches longer than the original to allow for subsequent trimming. On fitting, ensure the pipe is pushed fully onto its spigots to prevent fuel leakage.

5   Carburettor: removal and refitting

Note: *Petrol is extremely flammable, particularly in the form of vapour. Take all precautions to prevent the risk of fire and read the Safety first! section of this manual before starting work.*

1   Turn the fuel tap to the OFF position and disconnect the fuel pipe from the carburettor. Slacken the hose clamp which secures the air filter hose and the bolt(s) which secure the carburettor to the inlet stub. Pull the carburettor rearwards off the inlet stub and free it from the air filter hose.

2   Unscrew the threaded carburettor top and choke plunger retaining nut, withdraw the choke plunger and throttle valve, and remove the carburettor from the machine. It is not normally necessary to remove the throttle valve or choke plunger from their cables and they can be left attached and taped clear of the engine. However, if removal is necessary proceed as follows.

3   To remove the throttle valve, hold the carburettor top and compress the throttle return spring against, holding it in position against the top. Unhook the throttle cable and remove the valve and return spring. On the BVF 30 N 3-1 carburettor fitted to later 250 models, and all 251 and 300 models, the jet needle can be detached from the throttle valve by pushing down on the spring-loaded needle and sliding the needle retaining plate across the needle. Withdraw the needle, with its spring and spacer, through the circular cutout in the plate. Never remove the needle by pulling it through the slot as this will distort the retaining plate. On all other models, the needle and retaining plate can be simply tipped out of the throttle valve and the needle and plate separated.

4   The choke plunger is removed simply by sliding the cable out of the slot in the plunger and removing it along with its return spring and retaining nut.

5   Fitting the carburettor is a straightforward reversal of the removal procedure, whilst noting the following points. When refitting the throttle valve assembly into the carburettor body, ensure that the jet needle enters the needle jet smoothly and that the slot cut in the side of the throttle valve aligns with the pin in the carburettor body.

6   Ensure the carburettor is positioned vertically and tighten its retaining bolt(s) securely. Refit the air filter hose to the carburettor and tighten its retaining clamp securely. Adjust the idle speed as described in Section 8 of this Chapter, and if necessary, the oil pump cable as described in Routine maintenance. Thoroughly check for fuel leaks before taking the machine on the road.

5.1 Pull carburettor rearwards off inlet manifold and free it from air filter hose

5.2 Unscrew throttle valve and choke plunger and remove carburettor

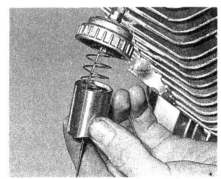

5.3a Compress return spring and disconnect throttle valve from the cable

5.3b To remove needle on 30 N 3-1 carburettor displace the retaining plate ...

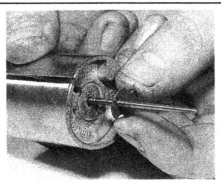

5.3c ... and withdraw needle through circular cutout

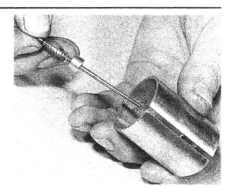

5.3d Needle assembly can then be withdrawn from top of throttle valve

5.3e On all other carburettors needle and retaining plate can be tipped out of valve once cable is disconnected

5.4 On refitting ensure slot in throttle valve aligns with pin in carburettor body

5.6 Inlet stub can be removed if necessary – check gasket

6 Carburettor: dismantling, examination and reassembly

Note: *Petrol is extremely flammable, particularly in the form of vapour. Take all precautions to prevent the risk of fire and read the Safety first! section of this manual before starting work.*

1   Before dismantling the carburettor, cover an area of the work surface with clean paper or rag. This will not only ensure that the carburettor components are kept clean, but will also prevent any of the smaller components being lost by making them more visible.

2   Release the float chamber by either removing its three retaining screws or wire retaining clip (as applicable), and remove the chamber and gasket from the carburettor body.

3   Gently tap out the float pivot pin and lift the float away from the carburettor. The float needle valve assembly can be unscrewed from the carburettor body using a socket or ring spanner. To dismantle the needle valve, use the spanner to retain the valve body, then unscrew the needle seat from top of the body and tip out the needle.

4   The main jet is a screw fit into the base of the needle jet, which is in turn a screw fit into the carburettor body. On later 250, 251 and 300 models (30 N 3-1 carburettor) the pilot jet is concealed behind a blanking screw situated next to the needle jet. Remove the screw and then unscrew the pilot jet. On all other models the pilot jet can simply be unscrewed from the carburettor. If necessary, the starter jet can be unscrewed from inside the float chamber.

5   Note the setting of the by-pass air or throttle stop screw (as applicable) by counting the number of turns required to screw it right in, then remove it from the carburettor. Record this information so that the screw can be returned to its original position on reassembly. Note that there is nothing to gain by unwarranted attention to these screws; only disturb if renewal or cleaning is required. Repeat the process for the pilot screw and remove it along with its spring. Note that on later 250, 251 and 300 models the pilot screw is concealed behind a plastic plug which

6.3a Gently tap out float pivot pin and lift float away

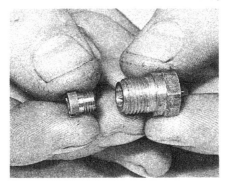

6.3b Unscrew needle seat from valve assembly ...

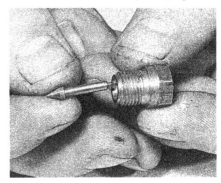

6.3c ... and tip out needle

Fig. 2.2 Carburettor – 125, 150 and early 250 models

1   Rubber cap – 2 off
2   Cable adjuster – 2 off
3   Locknut – 2 off
4   Carburettor top
5   O-ring
6   Return spring
7   Needle retaining plate
8   Jet needle
9   Throttle valve
10 Mounting clamp
11 Bolt
12 Washer
13 Nut
14 Carburettor body
15 Pilot jet
16 Needle jet
17 Main jet
18 Float
19 Gasket
20 Float chamber
21 Starter jet
22 Screw – 3 off
23 Spring – 2 off
24 Throttle stop screw
25 Pilot screw
26 Float pin
27 Choke plunger retaining nut
28 Return spring
29 Choke plunger
30 Plug
31 Float needle valve seat
32 Float needle
33 Float needle valve body

could be either white, red or black in colour. On these models gently tap the plug with a centre punch to displace it from the carburettor, then note the setting and remove the pilot screw and spring. Note that failure to note the settings of these screws will make it less easy to 'retune' the carburettor when it is reassembled and refitted to the machine.

6   On all models, thoroughly clean all components in clean petrol and examine them as follows.

7   After an extended period of service, the throttle valve will wear and may produce a clicking sound within the carburettor body. Wear will be evident on inspection, usually at the base of the valve and in the locating groove. A worn valve should be renewed as soon as possible as it will give rise to air leaks which will upset the carburation.

8   Examine the needle for wear and check that the needle is not bent

by rolling it along a flat surface such as a sheet of glass. If it is worn or bent it must be renewed together with the needle jet. On refitting ensure that the retaining plate is fitted to the specified groove of the needle (see Specifications).

9   Check the carburettor body, float chamber and the carburettor top carefully, looking for signs of damage such as cracks or splits, worn threads and distorted mating surfaces. Whilst it may be possible in some cases to repair such damage, it will usually be necessary to renew the damaged component.

10 Check that the float assembly is in good order and is not punctured. If either float is punctured, it will produce the wrong petrol level in the float chamber, leading to flooding and an over-rich mixture. If the floats are damaged, they must be renewed as it is not possible to effect

6.4a Main jet is a screw fit into base of needle jet ...

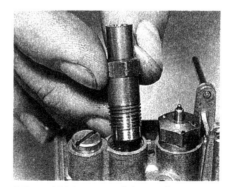

6.4b ... which is a screw fit into carburettor body

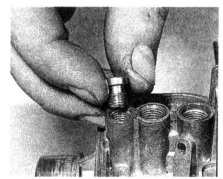

6.4c On 30 N 3-1 carburettor pilot jet is concealed behind threaded plug ...

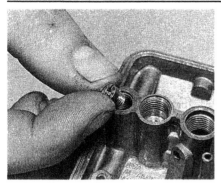

6.4d ... which must first be removed to gain access to it

6.4e Starter jet is located in float chamber

6.5 On 30 N 3-1 carburettor pilot screw is situated behind a plastic plug

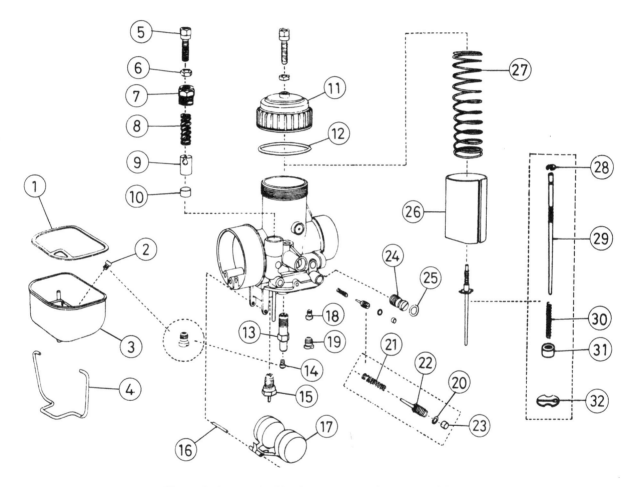

Fig. 2.3 Carburettor (30 N 3-1) – 251, 300 and later 250 models

1	Gasket	9	Choke plunger	17	Float	25	O-ring
2	Starter jet	10	Plug	18	Pilot jet	26	Throttle valve
3	Float chamber	11	Carburettor top	19	Blanking screw	27	Return spring
4	Retaining clip	12	O-ring	20	O-ring	28	Circlip
5	Choke cable adjuster	13	Needle jet	21	Spring	29	Jet needle
6	Locknut	14	Main jet	22	Pilot screw	30	Spring
7	Choke plunger retaining nut	15	Float needle valve	23	Plug	31	Spacer
8	Return spring	16	Float pin	24	Bypass air screw	32	Needle retaining plate

a satisfactory repair. Note that flooding could also be caused by an incorrect float height rather than a damaged float. If this is the case, check and adjust the float height as described in Section 8.

11   The float needle and seat will wear after lengthy service and should be closely examined with a magnifying glass. Wear will usually take the form of a ridge or groove, which will cause the float needle to seat imperfectly. If evident on either the seat or needle, the complete needle valve assembly must be renewed.

12   Examine the choke plunger for signs of wear or scoring, renewing it if necessary. The condition of its return spring can only be judged in comparison with a new component, and if there is any doubt about its condition it should be renewed. On later 250, 251 and 300 models renew

the O-rings which are fitted to the by-pass air and pilot screws, and on all models renew the O-ring fitted inside the carburettor top. If damaged, renew the float chamber gasket.

13   Before the carburettor is reassembled it should be cleaned out thoroughly using compressed air. Avoid using a piece of rag as there is always a risk of particles of lint obstructing the internal passageways or the jet orifices. Never use wire or any pointed object to clear a blocked jet. It is only too easy to enlarge the jet under these conditions and increase the rate of petrol consumption. If compressed air is not available, a blast of air from a tyre pump will usually be sufficient. As a last resort, a fine nylon bristle may be used.

14   Reassembly is basically a reversal of the removal procedure whilst noting the following points. Do not use excessive force as it is all too easy to shear one of the jets or small screws. If in doubt about the correct position of a component, refer either to the accompanying figure or photographs. When fitting the pilot screw and throttle stop or by-pass air screw (as applicable), ensure that each screw is screwed in until it seats lightly and then unscrewed to its previously noted position.

15   Before refitting the float chamber check, and adjust if necessary, the float height and fuel level as described in Section 8.

---

7 Carburettor: checking the settings

Note: *Petrol is extremely flammable, particularly in the form of vapour. Take all precautions to prevent the risk of fire and read the Safety first! section of this manual before starting work.*

1   The various jet sizes, throttle valve cutaway and needle position are predetermined by the manufacturer and should not require modification. Check with the Specifications for standard carburettor settings. If a change appears to be necessary it can often be attributed to a developing engine fault which is unconnected with the carburettor. Although carburettors do wear in service, this process occurs slowly over an extended period of time and hence wear in the carburettor is unlikely to cause sudden or extreme malfunction. If a fault does occur, check first other main systems, in which a fault may give similar symptoms, before proceeding with carburettor examination or modification.

2   Where non-standard items, such as an aftermarket exhaust system or air filter has been fitted, some alteration to the carburation may be needed. Arriving at the correct settings often requires trial and error, a method which demands skill borne of previous experience. In many cases the manufacturer of the non-standard equipment will also be able to advise on the carburation changes.

3   As a rough guide, up to $\frac{1}{8}$ throttle is controlled by the pilot jet, $\frac{1}{8}$ to $\frac{1}{4}$ by the throttle valve cutaway, $\frac{1}{4}$ to $\frac{3}{4}$ throttle by the needle position and $\frac{3}{4}$ to full by the main jet. These are only approximate divisions and are by no means clear cut. There is a certain amount of overlap between the various stages.

4   If alterations to the carburation must be made, always err on the side of a slightly rich mixture. A weak mixture will cause the engine to overheat which may cause engine seizure. Reference to Routine maintenance will show, how after some experience has been gained, the condition of the spark plug electrodes can be interpreted as a reliable guide to mixture strength.

---

8 Carburettor: adjustment

Note: *Petrol is extremely flammable, particularly in the form of vapour. Take all precautions to prevent the risk of fire and read the Safety first! section of this manual before starting work.*

1   Before any adjustment is made, eliminate all other possible causes of running problems, checking in particular the spark plug, contact breaker points, ignition timing, air filter and the exhaust. Checking and cleaning or adjusting these items will often resolve a mysterious flat spot or misfire.

2   The first step is to ensure that the jet sizes, needle position and float height are correct, which will require the removal and dismantling of the carburettor as described in Sections 5 and 6 of this Chapter.

Float height

3   If the carburettor has been removed for the purpose of checking jet

8.3 Checking float height

sizes, the float height should also be checked as follows. Remove the gasket from the sealing face of the carburettor and ensure that the surface is clean and smooth. Check that both floats are absolutely square to each other and that the bottom edge of each is at the same height from the carburettor surface. If necessary, realign the floats by carefully bending the bridge piece which joins the two.

4   Once the floats are known to be square and at the same height, hold the carburettor upright and measure the distance from the bottom edge of the floats to the sealing face of the carburettor. On 125 and 150 models this distance should be 33 + 1 mm (1.30 + 0.04 in), and on 250, 251 and 300 models it should be 32 mm (1.26 in). If necessary, this can be adjusted by bending the tang situated on the bridge piece of the floats which bears against the body of the carburettor.

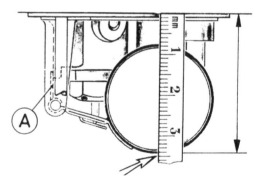

Carburettor upright – valve fully open

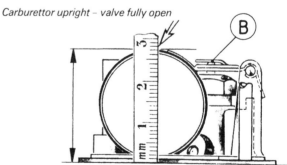

Carburettor inverted – valve closed

Fig. 2.4 Float height measurement (Sec 8)

A   *Stop lever*   B   *Float tang*

5   Invert the carburettor so that the floats are uppermost. The tang which bears on the float valve should be touching the valve, yet not compressing its spring-loaded tip. Measure the distance from the bottom of the floats to the sealing face of the carburettor. This distance should be 27 + 1 mm (1.06 + 0.04 in). If necessary, this distance can be adjusted by carefully bending the small tang which bears on the end of the float needle.

Fuel level
6   On all models the operation of the float and needle valve can be checked by measuring the fuel level as follows. To check the float level, a small clear petrol-proof container which will fit against the sealing face of the carburettor is needed to replace the float chamber. With the float heights set as described above, connect the carburettor to a fuel supply which is at least 0.5 m above the carburettor and hold the clear container tight against the sealing face of the carburettor. Ensure the carburettor is held vertically and switch the fuel supply on. Allow the fuel level in the container to settle then measure the distance from the top of the fuel to the sealing face of the carburettor. On 125 and 150 models this should be 12 ± 1 mm (0.47 ± 0.04 in), and 250, 251 and 300 models it should be 14 ± 1 mm (0.55 ± 0.04 in). If the float heights are known to be correct and the fuel level is incorrect, the float and needle valve should be removed for inspection as described in Section 6 of this Chapter.
7   Once the float settings and fuel level are known to be correct, refit the float chamber using a new gasket and refit the carburettor to the machine.

Adjustment screws
8   Start the engine and allow it to warm up to normal operating temperature, preferably by taking it for a short run. Then ensure there is 2.0 mm (0.08 in) of freeplay in the throttle cable and adjust the idle speed as described under the relevant sub heading.

Late 250, 251 and 300 models (30N 3-1 carburettor)
9   Remove the plastic plug, if not already having done so, which covers the pilot screw and screw the pilot screw in until it seats lightly. Then unscrew the pilot screw by three complete turns, start the engine, and adjust the by-pass air screw until the engine is running smoothly.

With the engine idling smoothly slowly screw in the pilot screw until the idle speed rises to its highest point, then back the pilot screw off by $\frac{1}{4}$ of a turn. If necessary lower the idle speed by backing off the by-pass air screw until the engine idle speed is 1200 ± 100 rpm, and refit the sealing plug over the pilot screw. MZ state that if the engine, ignition and exhaust system are in a good condition the pilot screw should be 2½ turns open and the by-pass air screw approximately 4 turns open.

All other models
10   Screw the pilot screw fully in until it seats lightly, then unscrew it by the number of turns shown in the Specifications at the start of this Chapter. Start the engine and adjust the throttle stop screw until the engine is idling at approximately 1200 rpm. Try turning the pilot screw in slowly by $\frac{1}{4}$ turn at a time, noting its effect on idle speed, then repeat the process, this time turning the screw outwards. Set the pilot screw to the position which gives the fastest consistent tickover, and if necessary lower the idle speed by unscrewing the throttle stop screw. Open and close the throttle a few times to check that the engine does not falter.

All models
11   Once the carburettor is correctly set, adjust the oil pump operating cable (where fitted) as described in Routine maintenance.

---

9   Air filter: general

1   The air filter element must be cleaned at the specified intervals, and renewed if damaged, to maintain good performance. Apart from the problem of increased wear caused by a damaged element, a clogged or broken filter will upset the carburation, allowing the mixture setting to become too rich or too weak.
2   Unless the air filter is damaged and obviously in need of renewal, it should require no attention other than the regular servicing described in Routine maintenance.
3   Never run the engine without the air filter element fitted as the carburettor is set up with the air filter fitted. If the element is removed, the mixture will become weak which will cause the engine to overheat and almost certainly lead to subsequent engine damage.

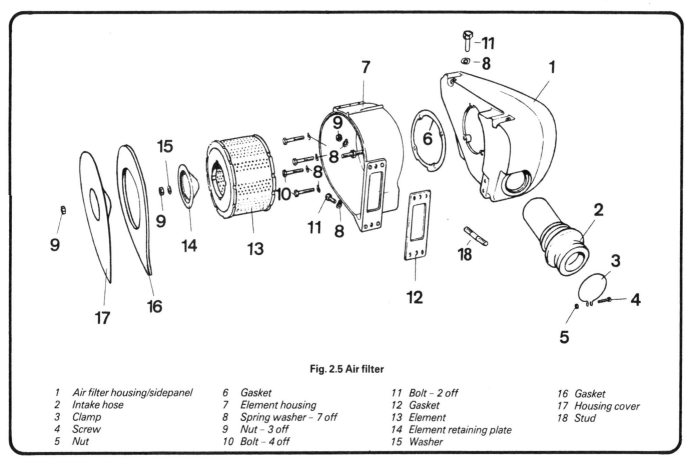

Fig. 2.5 Air filter

1 Air filter housing/sidepanel	6 Gasket	11 Bolt – 2 off	16 Gasket
2 Intake hose	7 Element housing	12 Gasket	17 Housing cover
3 Clamp	8 Spring washer – 7 off	13 Element	18 Stud
4 Screw	9 Nut – 3 off	14 Element retaining plate	
5 Nut	10 Bolt – 4 off	15 Washer	

10   Exhaust system: removal and refitting

1   The exhaust must be removed at regular intervals for the carbon deposits to be cleaned out; refer to Routine maintenance for details.

2   To remove the exhaust system, use a C-spanner to unscrew the ring nut which secures the exhaust pipe to the cylinder barrel, and then slacken and remove the bolt(s) at the silencer mounting clamp, and the bolt which secures the exhaust pipe/silencer clamp to the machine. The exhaust system can then be manoeuvred away from the machine. Take care not to lose the mounting rubbers from the silencer mounting clamp.

3   On refitting, locate the exhaust pipe in the cylinder barrel then refit the silencer mounting clamp rubber mountings and bolts and the bolt(s) which secures the exhaust pipe/silencer clamp to the machine. Tighten all the exhaust mountings by hand to settle it in position, then tighten the exhaust pipe ring nut to its specified torque setting and tighten all the mounting/clamp bolts securely.

4   Do not attempt to modify the exhaust system in any way. It is designed to give the maximum power possible and yet produce the minimum noise level possible. If an aftermarket system is being considered, check very carefully that it will maintain or increase performance when compared to the standard system, without making excessive noise.

11   Oil tank: removal and refitting – oil-injection models

1   To remove the oil tank it is first necessary to remove the complete air filter housing assembly. To do this remove the air filter element as described in Routine maintenance and slacken the clamp which secures the air filter hose to the carburettor. Remove the four bolts situated behind the element which retain the left-hand sidepanel/air filter housing assembly and manoeuvre it away from the machine.

2   The oil feed pipe should now be disconnected from the base of the oil tank. Before disconnecting the pipe prepare a short length of pipe of the same internal diameter as the original, which is plugged at one end. Also have ready a clean screw of suitable size to plug the end of the existing pipe once disconnected. These simple measures will prevent the loss of oil from the tank and seal the oil feed pipe while disconnected. Slide the feed pipe clip downwards off the tank stub. Very swiftly pull the feed pipe off the tank stub and replace it with the prepared pipe, plugged at one end. Use the screw to plug the end of the oil feed pipe.

3   Remove the two bolts from the rear of the tank, which secure it to its mounting bracket, and lift the tank away. Note the rubber dampers positioned between the tank and mounting bracket.

4   The tank is refitted by a reverse of the removal procedure. Examine the rubber dampers that are fitted between the tank and mounting bracket and renew them if they are perished or damaged. Offer up the

10.3a Locate exhaust pipe in cylinder barrel ...

10.3b ... and refit the rear mounting clamp

10.3c Tighten ring nut securely – to the specified torque setting if possible

10.3d Tighten all other mountings securely

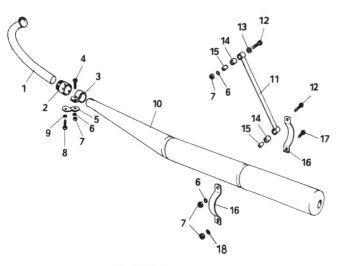

Fig. 2.6 Exhaust system

1 Exhaust pipe	10 Silencer
2 Ring nut	11 Silencer mounting rod
3 Clamp*	12 Bolt – 2 off
4 Bolt	13 Washer
5 Front mounting bracket*	14 Mounting rubber – 2 off
6 Spring washer – 3 off	15 Spacer – 2 off
7 Nut – 4 off	16 Rear mounting bracket
8 Bolt	17 Bolt
9 Spring washer	18 Spring washer

125 and 150 clamp assembly shown – other models similar

oil tank, positioning the rubber dampers between the tank and bracket, then refit the mounting bolts and washers and tighten them securely. Remove the screw from the oil feed pipe then quickly remove the sealed pipe from the oil tank and refit the feed pipe. Check the pipe is a tight fit on its stub and secure it with the retaining clip.

5   Position a new gasket on the sealing face of the left-hand side-panel/air filter housing assembly, refit it to the machine and tighten its retaining bolts securely. Reconnect the air filter hose to the carburettor and tighten its clamp securely. Refit the air filter element as described in Routine maintenance.

6   Bleed the oil pump as described in Section 13 of this Chapter before starting the engine.

12 Oil pump: removal and refitting – oil-injection models

1   The oil pump is mounted on the left-hand side of the crankcase and is accessed by removing the circular cover which is retained by two screws. The oil pump can be expected to give long service, requiring very little maintenance, but in the event of failure it must be renewed; there are no replacement parts available and the pump can therefore be considered a sealed unit.

2   To remove the pump, disconnect the oil feed and delivery pipes from the pump and plug their ends to prevent the escape of oil and the ingress of dirt; a clean screw or bolt of the appropriate diameter is ideal for this job. Disengage the oil pump control cable from the pump pulley and remove the two screws which secure the pump to the crankcase. The pump can then be lifted clear of the engine.

3   To refit the pump, clean both the pump and crankcase mating surfaces and place a new gasket on the pump. Align the driven spigot of the pump with the slot in the end of its driveshaft and fit the pump to the crankcase. Check the condition of the sealing washers which are fitted to the pump mounting screws, renewing them if necessary, then fit the screws and tighten them evenly and securely.

4   Unplug and swiftly reconnect both oil pipes. Ensuring that they are a good push fit on their respective stubs and secure them with their

12.1 Oil pump is located behind small circular cover on left-hand crankcase cover – 125 shown

12.3a Fit a new gasket to crankcase cover ...

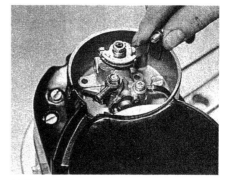

12.3b ... refit the pump ...

12.3c ... and tighten its mounting screws securely

12.4 Reconnect both oil pipes, securing them with their retaining clips ...

12.5 ... and connect oil pump pump operating cable

retaining clips. Examine the pipes for cracks and splits and check that they are routed correctly and are not twisted or kinked. Any fault found must be rectified immediately, as leakage will cause a loss of lubrication and could lead to engine seizure.

5   Reconnect the oil pump operating cable to the pump pulley and adjust it as described in Routine maintenance. Bleed the oil pump as described in the following Section before taking the machine on the road.

13   Oil pump: bleeding – oil-injection models

1   If the oil pump has been removed, or the oil tank run dry or drained, then it is most important to bleed any air from the system before the engine is started. Air in the system will rapidly produce an air lock which will interrupt the constant supply of oil, resulting in severe engine damage due to the consequent loss of lubrication.

2   Check that the oil pipes are correctly routed and are not trapped or kinked, and are held securely by their retaining clips. Check that the oil tank is full of oil, topping up to the filler neck if necessary. Thoroughly clean the oil pump and its surrounding area and pack clean rag around the base of the oil pump.

3   Remove the hexagon-headed bleed screw which is situated directly above the oil pump inlet union, and allow the oil to trickle out of the bleed hole whilst checking for air bubbles. Once the oil is flowing smoothly with no traces of air bubbles, refit the bleed screw and tighten it securely, and wipe away any trace of surplus oil.

4   If disturbed, it will also be necessary to prime the oil pump/crankcase delivery pipe. If this operation is not carried out the engine will be starved of oil until the pipe is filled. Start the engine and allow it to idle for a few minutes whilst holding the pump pulley in the fully open position. After a while the excess oil will make the exhaust start to smoke heavily, indicating that the oil delivery has reached the engine.

5   Once the bleeding process is complete, open and close the throttle

13.3 Oil pump bleed screw location

cable a few times to settle the oil pump cable, and check that it is properly adjusted as described in Routine maintenance.

14   Gearbox lubrication: general

1   General maintenance for the gearbox lubrication system consists solely of checking the level of the oil and changing it at the specified intervals. Carrying out these two service procedures will preclude any risk of the gearbox components being starved of oil or having to run in oil that has deteriorated to the point where it is ineffective as a lubricant.

2   Full details of checking the oil level and carrying out an oil change can be found in Routine maintenance.

Chapter 3 Ignition system

Refer to Chapter 7 for information relating to 1991-on models

Contents

Specifications

Ignition timing

Standard	2.5 mm (0.10 in) BTDC
Tolerance:	
300 models	2.5 – 2.7 mm (0.10 – 0.11 in) BTDC
All other models	2.5 – 3.0 mm (0.10 – 0.12 in) BTDC

Contact breaker gap

Standard	0.3 mm (0.012 in)
Tolerance	0.3 – 0.4 mm (0.012 – 0.016 in)

Spark plug

	Isolator	NGK
Recommended grade:*		
250 and 251 models	ZM 14-260	B7HS
All other models	ZM 14-260	B8HS
Electrode gap	0.6 mm (0.024 in)	

See notes in Routine maintenance on selection of spark plugs

1 General description

The ignition system is of the conventional battery and coil type, triggered by a contact breaker mounted on the generator stator, which is operated by a cam on the end of the crankshaft.

Opening and closing of the contact breaker every revolution, causes a magnetic field in the primary winding of the ignition coil to build up then collapse. This induces a high voltage in the secondary windings which causes the spark to occur across the electrodes of the spark plug.

2 Ignition system: fault diagnosis

1   Due to the nature of the ignition system, a fully charged battery is essential if the system is to operate normally. In the event of a fault in the ignition system, always check the battery and fuses before proceeding further.

2   Remove the spark plug, giving it a quick visual inspection for any obvious signs of flooding or oiling. Fit the plug into the suppressor cap and rest it on the cylinder head so that the metal body of the plug is in contact with the cylinder head. The electrode end of the plug should be positioned so that the sparking can be checked as the engine is spun over using the kickstart. Turn the ignition switch to the first position and kick the engine over. If the system is in a good condition a regular, fat blue spark should be evident at the electrodes. If the spark appears thin or yellowish, or is non-existent, further investigation is necessary.

3   The likely faults are listed below, starting with the most probable source of failure. Work through the list systematically, referring to the relevant Sections for full information of the necessary checks and tests.

(a)   Incorrect contact breaker gap. Check and adjust contact breakers as described in Routine maintenance

(b)   Incorrect ignition timing. Check and adjust as described in routine maintenance

(c)   Loose, corroded or damaged wiring connections, broken or shorted wiring between any of the ignition system components

(d)   Faulty condenser (capacitor)

(e)   Faulty ignition switch. Examine switch as described in Chapter 6.

(f)   Faulty ignition HT coil

3 Ignition system: checking the wiring

1   The wiring should be checked visually, noting any signs of corrosion around the various terminals and connectors. If the fault has developed in wet conditions it follows that water may have entered any of the connectors or the ignition switch, causing a short circuit. A temporary cure can be effected by spraying the relevant area with one of the proprietary water dispersant aerosols such as WD-40. A more permanent solution is to dismantle the switch or connector and coat the exposed parts with silicone grease to prevent the ingress of water. The exposed backs of the connectors can be sealed off using a silicone rubber sealant.

2   Light corrosion can normally be cured by scraping or sanding the

affected area, though in serious cases it may be necessary to renew the switch or connector affected. Check the wiring for chafing or breakage, particularly where it passes close to the fame or its fittings. As a temporary measure the damaged insulation can be repaired with PVC tape, but the wire concerned should be renewed at the earliest opportunity.

3   Using the wiring diagram at the end of this Manual, check each wire for breakage or short circuits using a multimeter set on the resistance scale or a dry battery and bulb arrangement as shown in Fig 6.1. In each case there should be continuity between the ends of each wire.

4   Condenser (capacitor): testing and renewal

1   The condenser prevents arcing across the contact breaker points as they separate; it is connected in parallel with the points. If misfiring occurs or starting proves difficult, particularly when the engine is hot, it is possible that the condenser is faulty. To check, watch the points as the engine is running. If the points spark excessively and appear burnt, the condenser requires renewal.

2   The condenser is located close to the contact breaker and is held by a strap, retained by a single screw. Before removing the condenser, detach the connecting link to the contact breaker by undoing the screw in the top of the condenser.

3   If the condenser malfunctions, no repair can be undertaken and a new component will have to be fitted. If malfunction is suspected, fit a new condenser in place of the original and observe the effect on engine performance. If the fault persists, it is most probably the ignition HT coil which is at fault.

5   Ignition HT coil: location and testing

The ignition HT coil is a sealed unit designed to give long service without the need for any regular maintenance other than ensuring that its mountings are tight and connections clean and secure. It is located underneath the seat. If a weak spark and difficult starting cause the performance of the coil to be suspect, it should be removed and taken to an authorised MZ service agent or a specialist auto-electrical expert, who will have the necessary equipment to test the coil. Resistance values for the primary and secondary windings are not available, with which it is normally possible to test the coil using home workshop equipment. A faulty coil must be renewed as it is not possible to effect a satisfactory repair.

6   HT lead and suppressor cap: examination and renovation

1   Erratic running faults can sometimes be attributed to leakage from the HT lead and spark plug cap, and it will often be possible to see tiny sparks around the lead and suppressor cap at night. One cause is dampness and the accumulation of road salts around the lead. It is often possible to cure the problem by drying and cleaning the components and spraying them with an aerosol ignition sealer. Water dispersant sprays such as WD-40 are also highly recommended where the system has become swamped with water. Both of these products can be obtained at most garages and accessory shops.

2   If the HT lead or suppressor cap are suspected of having broken down internally, they should be removed and renewed. The suppressor cap can be unscrewed from the end of the lead. To remove the HT lead from the ignition coil, unscrew its retaining nut and remove it along with the retaining sleeve. The HT lead can then be unscrewed from the coil.

4.1 Condenser (capacitor) is mounted on the stator housing

5.1 HT coil is situated on the right-hand side of rectifier

Examine both the HT lead cap seals and renew them if necessary.

3   On refitting, ensure that both ends of the lead are in a good condition, cutting the lead back to expose the copper core if necessary, and apply a small amount of silicone grease to each end of the lead to waterproof the connections. Screw the lead onto the coil, refit its retaining sleeve and nut and tighten it securely. Screw the suppressor cap onto the end of the lead and ensure both the sealing caps are correctly fitted.

7   Spark plug: general

If the spark plug is thought to be faulty, the only possible means of testing it is to renew the plug. Refer to Routine maintenance for further information.

Chapter 4 Frame and forks

Refer to Chapter 7 for information relating to 1991-on models

Contents

Specifications

Front forks

Travel	185 mm (7.4 in)
Fork spring free length:	
125 and 150 models	527 ± 4 mm (20.7 ± 0.16 in)
250, 251 and 300 models	527 mm (20.7 in)
Fork oil capacity:	
Normal	230 cc (8.09 fl oz)
Maximum:	
125 and 150 models	250 cc (8.79 fl oz)
250, 251 and 300 models	265 cc (9.32 fl oz)
Fork oil level – from bottom of fork:	
125 and 150 models:	
Normal	350 mm (13.78 in)
Maximum	370 mm (14.57 in)
250, 251 and 300 models:	
Normal	330 mm (12.99 in)
Maximum	395 mm (15.55 in)
Recommended fork oil	SAE10 fork oil

Rear suspension

Travel	105 mm (4.13 in)
Suspension unit spring free length:	
125 and 150 models	272 + 10 mm (10.71 + 0.39 in)
250, 251 and 300 models	260 + 8 mm (10.24 + 0.31 in)
Recommended damping oil	SAE10 fork oil

Torque settings

Component	kgf m	lbf ft
Steering stem nut:		
125 and 150 models	10.5 – 12.5	75.5 – 90
250, 251 and 300 models	15.0	108.0
Fork top bolts	15.0	108.0
Bottom yoke pinch bolts	2.0	14.5
Swinging arm pivot shaft nuts:		
125 and 150 models	8.0 – 10.0	57.5 – 72.0
250, 251 and 300 models	7.0 – 8.0	50.5 – 57.5
Tachometer drive gear – 250, 251 and 300 models	8.0 – 10.0	57.5 – 72.0

1   General description

The frame is of the spine type, using the engine as a stressed member.

Front suspension is by conventional coil sprung, hydraulically damped telescopic forks, and rear suspension is by two coil sprung hydraulically damped units which have 2 position spring preload adjusters. On 125 and 150 models, the performance of the rear suspension can be altered by changing the lower mounting point of both suspension units using the alternative mounting points provided in the swinging arm. The forwardmost holes provide the harder setting.

2   Front forks: removal

1   Remove the front wheel from the machine as described in Section 2 of Chapter 5. On models fitted with a front disc brake, remove the caliper mounting bolts, slide the caliper off the disc and position it clear off the fork legs. Tie the caliper to the frame to avoid placing any strain on its hydraulic hose. It is advisable to place a wooden wedge between the brake pads to prevent their movement if the brake lever is operated accidentally. On all models, slacken and remove the front mudguard mounting nuts and bolts and withdraw the mudguard from the machine.

2   Carefully lever off the plastic covers from the fork top bolts (where fitted), and slacken and remove the top bolts from the top of each stanchion. Slacken the bottom yoke pinch bolts and withdraw each leg by pulling it downwards out of the yokes. If corrosion hinders removal of the leg, apply a liberal quantity of penetrating fluid, allow time for it to work, then release the stanchion by rotating it in the yoke whilst pulling it downwards. In extreme cases, it is permissible to slightly open up the split clamp of the bottom yoke by fully removing the pinch bolt and working the flat blade of a screwdriver into the clamp. *Exercise extreme caution when doing this, as the clamps can be overstressed and are easily broken.*

3   If the fork legs are not to be dismantled, refit the top bolts to the stanchions to prevent the fork oil escaping or dirt entering the leg whilst they are stored.

3   Front forks: dismantling and reassembly

1   Dismantle and reassemble each fork leg separately to avoid interchanging components, thus producing undue fork wear.

2   Withdraw the fork spring from the stanchion noting which way up it is fitted. Invert the leg over a suitable container and pour out the fork oil, pumping the leg several times to expel as much oil as possible. Slacken the fork gaiter retaining clamp and remove the gaiter, or remove the dust cap (as applicable) from the top of the lower leg.

3   Unscrew and remove the damper rod retaining nut from the bottom of the lower leg, together with its washer. If the damper rod rotates, it will be necessary to use a flat-bladed screwdriver to hold the slotted end of the damper rod, whilst slackening its retaining nut with a box spanner.

4   Withdraw the stanchion assembly from the lower leg and remove the sealing washer, spring and stepped washer from the end of the damper rod, before tipping the damper rod out of the stanchion. Note that if the above components are not on the end of the damper rod as it is withdrawn, they will be inside the lower leg and should be tipped out.

5   Using a small flat-bladed screwdriver, carefully prise out the circlip from the lower end of the stanchion whilst pressing down on the damping valve it retains. Slowly release the damping valve and remove it from the stanchion along with the valve washer and spring. Behind the spring is another circlip which retains the spring seating washer. Remove the circlip and washer.

6   Remove the oil seal from the lower leg by levering it out, using a large flat-bladed screwdriver, whilst taking great care not to scratch or distort the top of the lower leg. If necessary, this task can be made easier by pouring boiling water over the top of the fork leg to expand the fork leg and release its grip on the seal, but take great care to prevent the risk of personal injury. Discard the oil seal; a new one should be obtained for reassembly.

7   Carefully clean all the fork components and examine them as

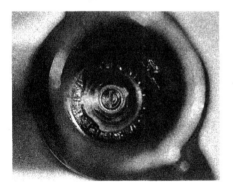

3.3 Remove damper rod retaining nut and withdraw stanchion from lower leg

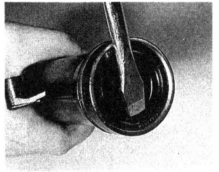

3.6 Take care not to damage lower leg when levering out fork seal

3.8a Tap fork seal into position using a suitable size tubular drift

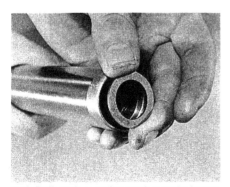

3.8b Refit spring seating washer to the lower end of stanchion ...

3.8c ... and secure it with its circlip

3.8d Refit the spring ensuring its large end is uppermost ...

3.8e ... followed by valve seat ,..

3.8f ... and damping valve and secure all components with second circlip

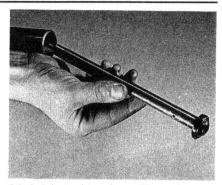

3.9a Lubricate piston ring and insert damper rod into stanchion upper end

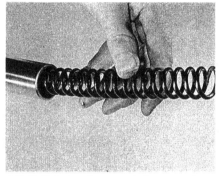

3.9b Refit fork spring to push damper rod out of lower end of stanchion ...

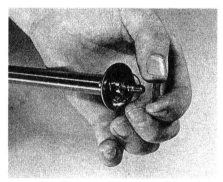

3.9c ... then fit stepped washer, spring and sealing washer to end of damper rod

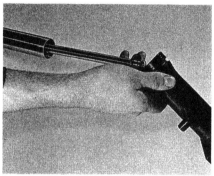

3.10 Lubricate fork seal then fit lower leg onto stanchion

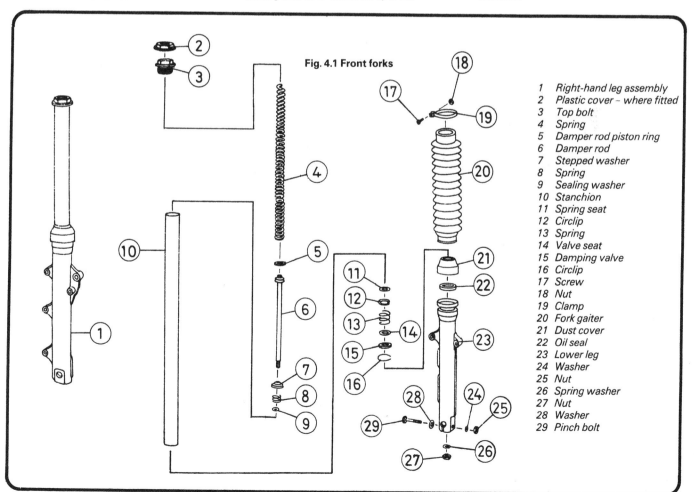

Fig. 4.1 Front forks

1   Right-hand leg assembly
2   Plastic cover – where fitted
3   Top bolt
4   Spring
5   Damper rod piston ring
6   Damper rod
7   Stepped washer
8   Spring
9   Sealing washer
10 Stanchion
11 Spring seat
12 Circlip
13 Spring
14 Valve seat
15 Damping valve
16 Circlip
17 Screw
18 Nut
19 Clamp
20 Fork gaiter
21 Dust cover
22 Oil seal
23 Lower leg
24 Washer
25 Nut
26 Spring washer
27 Nut
28 Washer
29 Pinch bolt

described in the following section. All traces of oil and swarf must be removed, and worn components renewed. Reassemble the fork leg as follows.

8   Apply a small amount of grease to the outer edge of the new oil seal and tap it into place using a suitably sized socket which bears only on the hard outer edge of the seal. Refit the spring seating washer to the lower end of the stanchion and secure it with its circlip. Then refit the spring with its larger end upwards, valve seat and damping valve and secure them with the second circlip. Ensure that both circlips are correctly seated in their grooves.

9   Lubricate the damper rod piston ring with clean fork oil and insert the damper rod into the upper end of the stanchion. Insert the fork spring into the upper end of the stanchion, ensuring that it is fitted the correct way around, and use it to push the damper rod fully into the stanchion. Use the spring to hold the damper rod in position and fit the stepped washer, spring and sealing washer to the end of the damper rod.

10   Lubricate the lip of the fork seal with clean fork oil and carefully insert the stanchion and damper rod assembly into the lower leg until the threaded end of the rod protrudes from the bottom of the lower leg. Refit the washer and nut which retain the damper rod, and tighten the nut securely. Refit the dust cover or gaiter (as applicable), ensuring that it is correctly seated on the lower fork leg, and refit the fork legs as described in Section 3 of this Chapter.

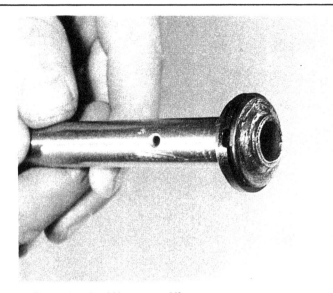

4.5 Piston ring should be renewed if suspect

use a fine file or emery paper to rub it down so that the original contour of the stanchion is restored. Damage of this kind can be eliminated by the fitment of fork gaiters on those models where they are not fitted already.

4   After an extended period of time, the fork springs may take a permanent set. If the spring lengths are suspect, then they should be measured and the readings obtained compared with those in the Specifications. If the free length of either spring has decreased by a significant amount, it is advised that they are renewed as a pair.

5   If the damping performance of the fork is suspect, even after an oil change, renew the damper rod piston ring and the damping valve.

6   Closely examine the dust seal or gaiter (as applicable) for signs of deterioration or splits. If found to be defective, it must be renewed as the ingress of any dirt will rapidly accelerate wear of the fork seal and stanchion. On models fitted with dust seals, it is strongly recommended that they be replaced with fork gaiters, as these offer much greater protection. These are available from any MZ dealer. Renew the fork oil seals as a matter of course, regardless of their apparent condition.

4   Front forks: examination and renovation

1   Examine the sliding surfaces of the stanchion and the internal surfaces of the lower leg for signs of wear or damage, which will be visible as scuffing. The degree of wear can be checked by inserting the stanchion fully into the lower leg and moving the stanchion backwards and forwards and from side to side. Although it is inevitable that there will be a certain amount of movement, an excessive amount of freeplay will indicate that both the stanchion and lower leg require renewal. However, as no figures are available, it is largely a matter of experience to assess with accuracy the amount of wear necessary to justify the renewal of either component. If in doubt, it is recommended that both components are taken to an authorized MZ dealer for inspection.

2   Check the outer surface of the stanchion for scratches or roughness; it is only too easy to damage the oil seal if these high spots are not eased down. The stanchions are unlikely to be bent unless the machine has been involved in an accident. Any significant bend will be detected by eye, but if there is any doubt about straightness, roll the stanchion tubes along a flat surface. Unless specialised repair equipment is available it is rarely practicable to effect a satisfactory repair. Although very slightly bent stanchions may be straightened, it should be noted that MZ specifically advise against this.

3   Check the stanchion surface for pits caused by corrosion or stone chips. Such pits should be smoothed down using fine emery paper and filled, if necessary, with Araldite. Once the Araldite has set fully hard,

5   Front forks: refitting

1   Remove all traces of dirt and corrosion from the bores of the yokes, then apply a smear of grease to aid refitting of the fork legs. Thoroughly clean the fork stanchions and insert the fork legs into the yokes. Position the legs so that the top of each stanchion is tight against the top yoke

5.1 Refit stanchions to the yokes and refill with specified amount of oil

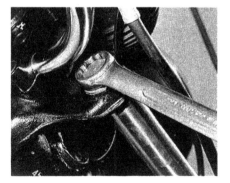

5.3a First tighten both fork top bolts to specified torque setting ...

5.3b ... followed by bottom yoke pinch bolts

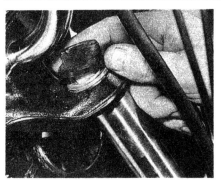

5.3c Refit the plastic covers (where fitted) to fork top bolts ...

5.3d ... and slide gaiter up to bottom yoke – tighten its clamp securely

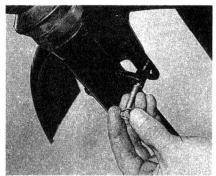

5.4 Refit the mudguard mounting bolts but do not tighten them until front wheel has been fitted

and temporarily tighten the bottom yoke pinch bolts.

2   If the fork legs were dismantled, fill each leg with the correct amount of the specified fork oil and then slowly pump the lower leg to distribute the oil around the fork components. On all models, whether the fork legs were dismantled or not, check the oil level as described in Routine maintenance.

3   Once the fork oil level is correct, apply a thin smear of jointing compound to the threads of the top bolts and refit them to the stanchions. Slacken the bottom yoke pinch bolts and tighten both top bolts to the specified torque setting, and refit the plastic covers (where fitted). Ensure that the turn signal lamps are correctly aligned and tighten the bottom yoke pinch bolts to the specified torque setting. Slide the fork gaiters (where fitted) up to the bottom of the bottom yoke and tighten their retaining clamps securely.

4   Refit the mudguard, tightening its mounting bolts finger-tight only at this stage, and refit the brake caliper to the lower fork leg (having removed the wooden wedges), tightening its mounting bolts securely.

5   Refit the front wheel as described in Section 3 of Chapter 5, tightening its spindle and pinch bolt to their specified torque settings, then tighten the mudguard mounting bolts securely. Finally, thoroughly check the operation of the front forks and brake before taking the machine out on the road.

6   Steering head: removal

1   Remove the fuel tank as described in Chapter 2. Remove the fork legs as described in Section 2 of this Chapter.

2   Disengage the rubber cover(s) from the instrument(s) then lift out the instrument(s) and disconnect the instrument drive cable(s). Remove the cover from the bottom of the ignition switch, make a note of how the relevant instrument and ignition switch wiring is connected and disconnect all wires. Prise off the four black plastic caps fitted to the handlebar clamp bolts, remove the handlebar bolts and lift off the instrument panel/handlebar clamp as an assembly.

3   Move the handlebars to the rear of the top yoke so they are clear of the steering head area. Take care not to stretch, trap or damage the control cables or wiring, and secure the handlebars to the frame. On disc brake models do not place any strain on the hydraulic hose and position the bars so that the master cylinder reservoir is upright to avoid the possibility of brake fluid spillage.

4   Remove the screw situated at the bottom of the headlamp rim, which secures the headlamp unit to its shell and partially remove the headlamp unit. Disconnect the headlamp and parking lamp wiring and remove the headlamp from the machine. Slacken the large nut which secures the headlamp shell to the bottom yoke and disengage the shell from the yoke. The shell can be left in position and, if necessary, tied to the frame for support. Remove the bottom yoke pinch bolts and tie the turn signal lamps to the frame to avoid straining their wiring connections.

5   Prise the plastic cap off the steering stem nut and slacken the nut. Remove the steering stem nut, whilst supporting the bottom yoke, then lift off the top yoke and lower the bottom yoke out of the steering head.

If necessary, the top yoke can be tapped off of the steering stem using a soft-faced mallet.

6   The bearings can be removed as follows. *Do not remove the steering head bearings unnecessarily; they will almost certainly be damaged during the removal process.* Pass a long drift down through the steering head and move the central spacer to one side until the inner race of the lower bearing can be reached. Moving the drift evenly around its inner race, drive the lower bearing out of position with a few blows from a heavy hammer. Withdraw the central spacer, then drift out the upper bearing from below the steering head. Examine the bearings and central spacer as described in the following Section.

7   Steering head bearings: examination

1   If the regular check described in Routine maintenance reveals bearing play or damage, the bearings must be removed as described in the previous section and checked as follows.

2   Wash the bearings in a high flash-point solvent to remove all traces of old grease, dry them and check for wear as follows. Hold the inner race of the bearing firmly and check for freeplay at the outer race. If free play is present the bearing is worn. Spin the outer race of the bearing and check that it spins quietly and slows down smoothly. If the bearing rotates noisily or slows down jerkily, it is worn. Also examine the bearing visually for signs of damage such as damaged balls, bearing tracks and cages. If either bearing shows signs of wear or damage, renew both bearings as a set.

3   Examine the central spacer for signs of wear or damage, and measure the length of the spacer. If the spacer shows signs of wear or damage, or is less than 171 mm (6.73 in) in length, it must be renewed. If a spacer of less than the required length is used, the bearings will be subjected to stress and will wear prematurely.

8   Steering head: refitting

1   If the steering head bearings have been removed, thoroughly clean the inside of the steering head with a high flash-point solvent to remove all traces of old grease, and smooth away all burrs or raised edges which may hinder bearing installation. Pack the bearings with high melting-point grease and smear grease over both sides of the central spacer and on the inside of the steering head. This will help prevent corrosion and aid refitting.

2   The bearings can be fitted using a drawbolt arrangement (see accompanying figure). Firstly obtain a large high-tensile bolt which is long enough to pass through the steering head, both bearings and the two washers and have the nut fitted to its end. Ideally, it should also be sufficiently large in diameter to fit closely through the bearing inner races, so that it can locate the central spacer in the correct position at the

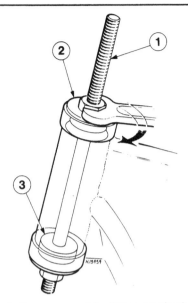

Fig. 4.2 Drawbolt arrangement for fitting steering head bearings (Sec 8)

1   *Drawbolt*   3   *Guide*
2   *Thick washer*

same time. Two thick washers which are of the same diameter as the outer races of the bearings, and are a close fit over the bolt will also be needed.

3   Fit a thick washer to the bolt followed by a bearing, ensuring that its sealed face is facing outwards, and the central spacer. Pass the bolt up through the steering head and fit the second bearing to the bolt, again ensuring its sealed surface is facing outwards, followed by the second thick washer and nut; tighten the bolt finger-tight only. Position the bearings in the steering head and slowly pull them into position by tightening the nut and bolt, whilst ensuring the bearings remain square. When both bearings are correctly seated, unscrew the nut and remove the bolt and washers. Check that the central spacer is correctly aligned, levering it into position if necessary.

4   Smear the steering stem with grease and insert the bottom yoke assembly into the steering head. Refit the top yoke and steering stem nut, and tighten the nut by hand only at this stage.

5   Refit the handlebars, ensuring that all cables and wiring is correctly routed, then refit the instrument panel/handlebar clamp assembly. Tighten its mounting bolts securely and refit the plastic caps. Reconnect the instrument panel and ignition switch wiring, using the notes made on dismantling, and refit the cover to the bottom of the ignition switch. Reconnect the speedometer drive cable and slide the instrument into position, ensuring it is correctly seated in the rubber cover, then repeat the process for the tachometer (where fitted). Refit the turn signal lamps to the bottom yokes, tightening the pinch bolts finger tight only.

6   Install the fork legs to align the top and bottom yoke, and tighten the steering stem nut to its specified torque setting. Check that the yokes move smoothly from lock to lock with no sign of free play, and refit the stem nut cover. Refit the mudguard and front wheel and tighten all fork mounting bolts as described in Section 5.

7   Refit the headlamp shell to the bottom yoke, remake the wiring connections and refit the headlamp unit. Adjust the headlamp beam as described in Routine maintenance and tighten the headlamp shell mounting nut securely. Finally, check that the handlebars move smoothly from lock to lock, ensuring that all control cables and wiring leads are correctly routed and in no danger of becoming trapped. Check the operation of all the disturbed electrical components before taking the machine on the road.

9   Frame: examination and renovation

1   The frame is unlikely to require attention unless accident damage

has occurred. In some cases, renewal of the frame is the only satisfactory remedy if it is badly out of alignment. Only a few frame specialists have the jigs and mandrels necessary for resetting the frame to the required standard of accuracy, and even then there is no easy means of assessing to what extent the frame may have been overstressed.

2   After the machine has covered a considerable mileage, it is advisable to examine the frame closely for signs of cracking or splitting at the welded joints. Rust corrosion can also cause weakness at these joints. Minor damage can be repaired by welding or brazing, depending on the extent and nature of the damage.

3   Remember that a frame which is out of alignment will cause handling problems and may even promote 'speed wobbles'. If misalignment is suspected, as a result of an accident, it will be necessary to strip the machine completely so that the frame can be checked, and if necessary, renewed.

10   Swinging arm: removal

1   Remove the rear wheel and the sprocket assembly as described in Sections 4 and 7 of Chapter 5. Slacken and remove both the upper and lower suspension unit mounting bolts and remove both units from the machine.

2   Unscrew the locknut from the left-hand end of the swinging arm pivot shaft, and remove the left-hand pivot shaft nut. Then using a hammer and suitable drift, tap the pivot shaft out of position and withdraw the swinging arm from the machine, noting the thrust washers which are fitted between the swinging arm pivot lugs and the frame.

11   Swinging arm: examination and renovation

1   Thoroughly clean all components, removing all traces of dirt, grease and corrosion.

2   Inspect the swinging arm closely for signs of cracks or distortion, although such damage is only likely if the machine has been involved in an accident. If any such damage is discovered, the swinging arm must be renewed. If its painted finish has deteriorated it is worth taking the opportunity to repaint the affected area, ensuring that the surface is correctly prepared beforehand.

3   Check the pivot shaft for signs of wear along the length of its shank. If the shank is seen to be stepped or badly scored, it must be renewed. Check the shaft for straightness by rolling it on a flat surface such as a sheet of glass. If the shaft is not perfectly straight it must be renewed. Also ensure that its threads and those of its retaining nuts are in a good condition. Examine the thrust washers, positioned between the swinging arm pivot lugs and the frame, for signs of wear or damage and renew as necessary.

4   The rubber bushes are an interference fit in the swinging arm pivot lugs, and are unlikely to wear or deteriorate until a high mileage has been covered. However, wear may occur between the metal inner sleeves and the pivot shaft, especially if either retaining nut has become loose. In normal service there should be no movement between the inner sleeves and pivot shaft, as all the swinging arm movement is allowed for in the flexing of the rubber bush. Examine the rubber bushes for signs of damage or deterioration, and examine the inner surface of the metal inner sleeves for signs of scoring. If either component shows signs of damage or wear, both the bushes and inner sleeves should be renewed as a set.

5   Note that while it may be possible to drive or press out the inner sleeve and cut out the rubber bush, the new components will be an extremely tight fit in the swinging arm lugs and fitting will prove extremely difficult without the necessary service tools. Fitting the new bushes will require the use of a press a special stepped mandrel to the guide the bush so that it is not distorted as it is fitted, while fitting the inner sleeves requires the use of the press again and a tapered, stepped mandrel which can be fitted inside the sleeve and will guide it into the bush. Therefore, if the swinging arm bushes and sleeves require

Fig. 4.3 Swinging arm and rear suspension (Sec 10)

1   Pivot shaft nut – 2 off
2   Pivot shaft
3   Inner sleeve – 2 off
4   Bush – 2 off
5   Swinging arm
6   Washer
7   Washer
8   Pivot shaft locknut
9   Plug – 2 off
10 Chain adjuster – 2 off
11 Washer
12 Nut – 4 off
13 Suspension unit damper – 2 off
14 Preload adjuster – 2 off
15 Spring – 2 off
16 Protective sleeve – 2 off
17 Split collets – 2 off
18 Lower mounting bolt – 2 off
19 Spring washer – 2 off
20 Washer – 2 off
21 Spacer – 2 off
22 Mounting rubber – 2 off
23 Upper mounting bolt – 2 off
24 Washer – 2 off
25 Bush – 2 off

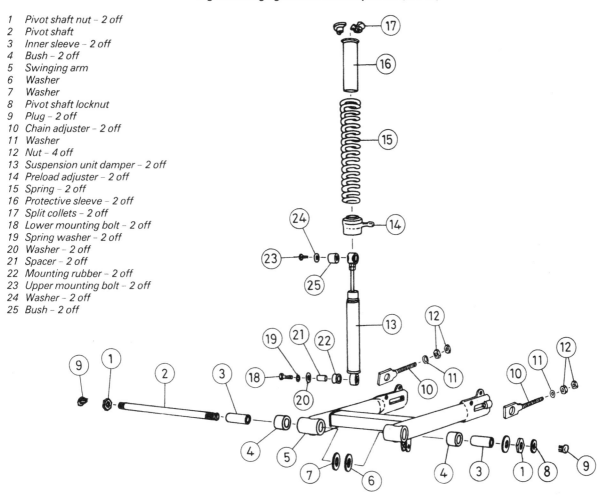

11.4 Examine rubber bushes for signs of wear or deterioration and renew if necessary

renewal it is highly recommended that the job be entrusted to an authorized MZ dealer, who will have the necessary equipment.

12 Swinging arm: refitting

1   Thoroughly grease the pivot shaft, the swinging arm lug inner sleeves and the bore of the frame through which the pivot shaft passes.
2   Offer up the swinging arm, fit the thrust washers between the swinging arm pivot lugs and the frame and insert the pivot shaft from the right-hand side. Refit the left-hand pivot shaft nut and tighten it by hand only.
3   Refit the rear suspension units to the machine, tightening their mounting bolts securely, then refit the rear sprocket assembly and rear wheel as described in Sections 5 and 7 of Chapter 5.
4   Push the machine off its centre stand and bounce the rear suspension several times to settle the components. Then with the rider seated normally on the machine, tighten the left-hand pivot shaft nut to the specified torque setting. It is essential that the swinging arm is settled in its normal position when the pivot shaft is tightened to clamp the inner sleeves as this minimises the distortion of the rubber bushes and will greatly prolong their life. Finally, refit the locknut and tighten it securely.

13 Rear suspension units: examination and renovation

1   To remove the suspension units, unscrew both the upper and lower

12.2 Do not omit thrust washers when refitting swinging arm

12.3a Ensure both upper ...

12.3b ... and lower suspension unit bolts are tightened securely

12.4 Tighten pivot shaft nut as described in text then tighten locknut securely

suspension unit mounting bolts and lift the units clear of the machine.

2   Clamp the lower suspension unit mounting lug in a vice equipped with soft jaws and set the spring preload adjuster to the softest setting. With the aid of an assistant, compress the suspension unit spring and remove the split collets from the top of the unit. The protective sleeve, spring and preload adjuster can then be removed from the unit. Repeat the process for the second unit.

3   Examine the units for signs of fluid leakage. If there is any sign of leakage on either unit, both damper units should be renewed as a pair. If there is no signs of leakage, check the operation of each unit by pushing

and pulling the damper rod in and out of the unit whilst holding it in a vertical position. An even resistance should be felt in both units with no sudden easy movements. If this is not the case, both units should be dismantled and overhauled as follows.

4   The damper units are dismantled by unscrewing the threaded ring from the top of the oil reservoir. To unscrew this, it will be necessary to use the MZ service tool Part Number 05-MW 82-4, or fabricate a homemade spanner for the task. Once the ring has been slackened, withdraw the ring along with the damper rod assembly and tip out the old oil into a suitable container. Clean all damper unit components in solvent and dry them completely. Examine all components for signs of wear or damage, noting that although the units can be dismantled, no spare parts are available. If any component is found to be faulty, both damper units must be renewed as a pair. If all appears well, top up the oil reservoir with the specified oil, refit the damper rod assembly and tighten the threaded ring securely. If possible, the ring should be tightened to approximately 5.0 kgf m (36 lbf ft). Recheck the operation of both damper units.

5   Examine the springs for signs of cracks or damage. The amount of wear can be assessed by measuring the free length of the springs. If the free length of either spring is less than the service limit given in the Specifications, or there is a considerable difference between the length of the two springs, both springs should be renewed as a pair.

6   Each suspension unit is fitted with a rubber bush in each of its mounting lugs. Examine the bushes for signs of wear or deterioration and renew them as necessary. The bushes can be pressed out using a vice and two sockets, one large one to support the damper unit, and a smaller one with the same diameter as the bush as a mandrel.

7   On reassembly, refit the preload adjuster, spring and protective sleeve to the damper unit. Ensure that the preload adjuster is set to the softest position and with the help of an assistant, compress the spring and refit the split collets to the top of the unit. Check that the collets are correctly seated, refit the units to the machine and tighten their mounting bolts securely. Set the preload adjusters to the required position and thoroughly check the operation of the rear suspension before taking the machine on the road.

13.2a Compress spring, remove split collets from top of suspension unit ...

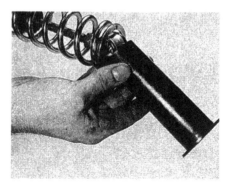

13.2b ... and slide off protective sleeve, spring ...

13.2c ... and preload adjuster

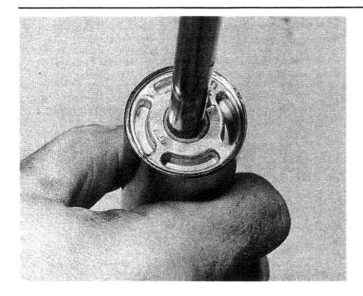

13.4 Threaded ring can be unscrewed using a suitable peg spanner

14 Footrests, brake pedal and stand: examination and renovation

1   The rider's footrests are secured to the frame by either two or three bolts, depending on the model. To remove the rear brake pedal, first mark the left-hand end of the pedal shaft and operating linkage arm to use as a reference on refitting, then remove the pinch bolt and disengage the arm from the shaft and return spring. The brake pedal can then be withdrawn from the right-hand side of the machine and the return spring removed from the left-hand side of the frame. To remove the stand, lean the machine against a wall and remove the circlip fitted to the right-hand side of the stand and its return spring. Move the stand first to the right and then to the left to disengage it from its pivots, and remove it from the machine.
2   Thoroughly clean all components, removing all traces of old grease, road dirt and corrosion and examine them for wear and damage. Check also that there is no damage such as bent cracked or split components, although such damage is unlikely unless the machine has been involved in an accident. Check the stand and brake pedal return springs for signs of wear or damage and renew them if there is the slightest doubt about their condition.
3   All steel components can be easily repaired by welding or brazing if cracked or broken. If bent in an accident, the component should be heated to a dull cherry red colour before being straightened. In either case the surface of the repaired component will require repainting afterwards. Repairs may be possible to the alloy centre stand, although such repairs are strictly for the expert only. Refit the components as follows.
4   Apply a small amount of grease to the stand pivot points and refit the stand, engaging it first with its left-hand pivot and then the right-hand. Refit its retaining circlip to the right-hand pivot point and fit the return spring. Smear a small amount of grease along the brake pedal shaft and insert the pedal into the frame. Fit the return spring to the left-hand side of the frame and refit the operating arm to the brake pedal shaft splines using the notes made on dismantling to position it correctly. Ensure the operating arm is correctly engaged with the return spring, then refit its pinch bolt and tighten it securely. Offer up the footrests, refit the mounting bolts and tighten them securely.
5   Before taking the machine on the road, ensure that the footrests are securely mounted, and check that the rear brake pedal is operating correctly. Also ensure that the stand return spring holds the stand securely in the retracted position.
6   The pillion footrests are secured by nuts and spring washers to brackets extended from the frame and can be easily renewed if damaged. They require no maintenance except for a few drops of oil at regular intervals.

15 Speedometer and tachometer drives: examination and renovation

Cables
1   If an instrument suddenly fails to operate or becomes sluggish, remove the cable from the machine and check that it is not broken or dry through lack of lubricant. The speedometer and tachometer (where fitted) drive cables are secured at their upper and lower ends by knurled retaining rings. These rings should be slackened and tightened using pliers.
2   Withdraw the inner cable and examine both the inner and outer cables for signs of damage. Do not check the inner cable for broken strands by passing it through the fingers or palm of your hand as this will cause a painful injury if a broken strand snags the skin. The best way to check the inner cable is to wrap a piece of rag around it and pull it through the rag, any broken strands will then snag the rag. If either the inner or outer cable is damaged, the complete cable must be renewed. If the old cable is to be re-used, grease the cable as described in Routine maintenance before refitting it.
3   On refitting ensure each cable is correctly routed with no kinks or sharp bends, and that they are secured by any guides, clamps or ties provided. Turn the rear wheel (or turn the engine over on the kickstart, as applicable) to help the drive mechanism engage with the inner cable and tighten the retaining rings securely.

Speedometer drive
4   The speedometer drive is situated in the rear sprocket cover and will rarely give any trouble. However, if suspect it can be tested by disconnecting the cable from the drive and spinning the rear wheel. If the drive pinion fails to rotate, the speedometer drive should be dismantled and checked as described in Section 7 of Chapter 5.

Tachometer drive
5   On 125 and 150 models the drive mechanism is inside the left-hand crankcase cover, where it is situated behind the clutch assembly which must first be removed to gain access, whereas on 250, 251 and 300 models it is mounted on the outside of the left-hand crankcase cover and can be removed without any preliminary dismantling.
6   To test the tachometer drive, disconnect the cable from the engine and turn the engine over on the kickstart. If the drive pinion fails to rotate, the tachometer drive should be removed for examination. On 125 and 150 models the tachometer drive can be dismantled as described in Chapter 1, and on 250, 251 and 300 models it should be checked as follows.
7   Remove the three screws which secure the tachometer drive housing to the cover and remove it from the machine. Remove the screw and washer from inside the housing, and withdraw the threaded bush and driven gear. To remove the drive gear from the end of the crankshaft, select top gear and apply the rear brake (rear wheel touching the ground) to prevent the crankshaft from rotating as the drive gear is unscrewed.
8   Thoroughly clean all components and examine both gears for signs of chipped or broken teeth. If either gear is damaged, renew both gears as a pair. Examine the internal surface and the threads of the plastic bush, renewing it if damaged in any way.
9   On reassembly smear the shaft of the driven gear with grease, then fit the washer and plastic bush. Insert the gear and bush into the tachometer housing, then refit the retaining screw and washer and tighten it securely. Fit the tachometer drive gear, preventing the crankshaft from rotating as described above, and tighten it to the specified torque setting. Refit the housing assembly to the casing, ensuring that the tachometer gears engage correctly, and tighten its retaining screws securely.

16 Speedometer and tachometer heads: removal, examination and refitting

1   To remove the speedometer or tachometer (as applicable), disengage the rubber instrument cover, then lift out the instrument and disconnect its drive cable. Pull out all the bulbholders from the bottom of the instrument, noting where each was fitted, and then lift the instrument away from the machine.

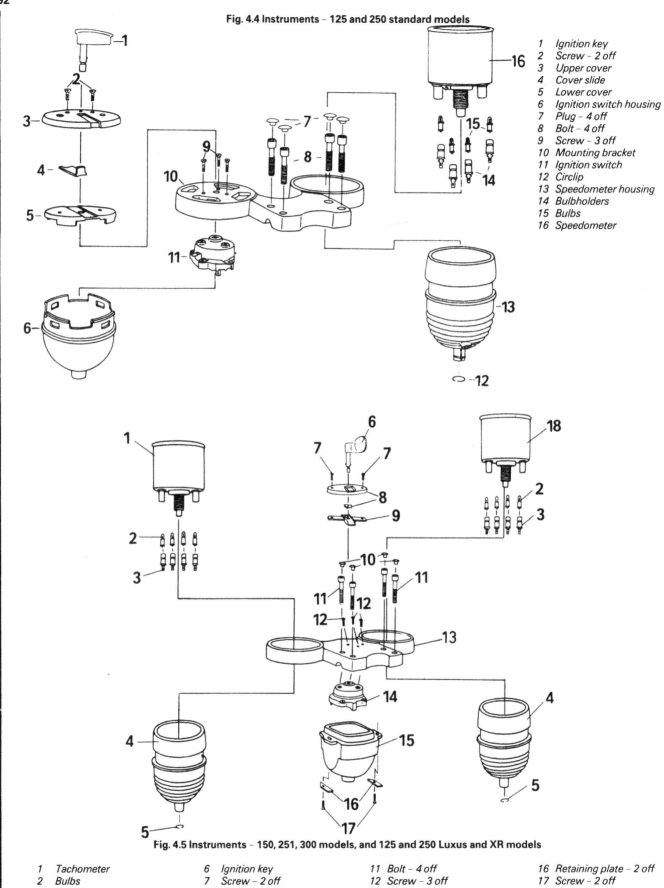

Fig. 4.4 Instruments – 125 and 250 standard models

1 Ignition key
2 Screw – 2 off
3 Upper cover
4 Cover slide
5 Lower cover
6 Ignition switch housing
7 Plug – 4 off
8 Bolt – 4 off
9 Screw – 3 off
10 Mounting bracket
11 Ignition switch
12 Circlip
13 Speedometer housing
14 Bulbholders
15 Bulbs
16 Speedometer

Fig. 4.5 Instruments – 150, 251, 300 models, and 125 and 250 Luxus and XR models

1	Tachometer	6	Ignition key	11	Bolt – 4 off	16 Retaining plate – 2 off
2	Bulbs	7	Screw – 2 off	12	Screw – 3 off	17 Screw – 2 off
3	Bulbholders	8	Upper cover and slide	13	Mounting bracket	18 Speedometer
4	Instruments housing – 2 off	9	Lower cover	14	Ignition switch	
5	Circlip – 2 off	10	Plug – 4 off	15	Ignition switch housing	

16.1a To remove instrument head, disengae it from its rubber cover ...

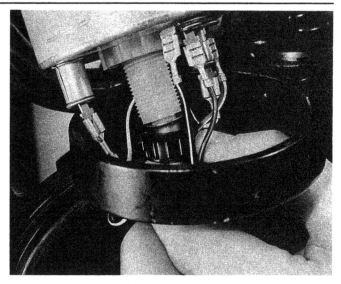

16.1b ... then disconnect drive cable and remove all bulbholders

2　The instruments must be carefully handled at all times and must never be dropped or held upside down. Dirt, oil, grease and water all have an equally adverse effect on the instrument heads.

3　The instrument heads are very delicate and should not be dismantled at home. In the event of a fault developing, an instrument should be entrusted to a specialist repairer or a new unit fitted. If a replacement unit is required, it is well worth trying to obtain a good secondhand item from a motorcycle breaker in view of the high cost of a new one.

4　Remember that a speedometer in a correct working order is a statutory requirement in the UK. Apart from this legal necessity, reference to the odometer readings is the most satisfactory means of keeping pace with the maintenance schedules.

5　On refitting, ensure the bulbholders are fitted to their original positions in the instrument and connect the drive cable. Refit the instrument to the rubber cover, ensuring that the two locate and seal correctly.

Chapter 5 Wheels, brakes and tyres

Refer to Chapter 7 for information relating to 1991-on models

Contents

Specifications

Wheels

Rim size:
Front	1.60 x 18
Rear:	
125 and 150 models	1.85 x 16
250 and 300 models	2.15 x 18
251 models	2.15 x 16

Front disc brake

Disc diameter	280 mm (11.0 in)
Disc standard thickness	4.9 – 5.2 mm (0.193 – 0.205 in)
Service limit	4.5 mm (0.177 in)
Maximum difference	0.025 mm (0.001 in)
Disc standard runout	Less than 0.2 mm (0.008 in)
Service limit	0.3 mm (0.012 in)

Drum brake – front and rear

Drum diameter:
125 and 150 models	150 mm (5.9 in)
250, 251 and 300 models	160 mm (6.3 in)
Shoe width – all models	30 mm (1.2 in)

Tyres

Size:
Front	2.75 x 18
Rear:	
125 and 150 models	3.25 x 16
250 and 300 models	3.50 x 18
251 models	3.25 x 16 or 110/80 x 16

Tyre pressures – tyres cold

Tyre pressures – tyres cold .... See Routine maintenance specifications

Torque settings

Component	kgf m	lbf ft
Front wheel spindle	8.0	58.0
Front wheel spindle pinch bolt	2.0	14.5

1 General description

The wheel rims are of the light alloy type and are laced to the hub by steel spokes. The rear wheel can be removed from the machine without disturbing the chain since the rear sprocket rotates on a separate spindle. A large rubber cush drive is fitted between the rear wheel and sprocket to minimise transmission shock loadings.

Front brake is by either an hydraulically operated disc or a cable operated drum brake. The disc brake is of the opposed piston type, and the drum brake is of the single leading shoe type. On all models the rear brake is of the single leading shoe drum type.

The tyres are of the conventional tubed type.

2 Front wheel: removal

1 Place the machine on its stand and place a support underneath the crankcase to prevent the machine from falling forwards when the front wheel is removed.

2 On models fitted with a front disc brake remove the wheel spindle nut and slacken the spindle pinch bolt. Remove the spindle nut and washer and withdraw the spindle. If necessary the spindle can be tapped out of position using a hammer and a suitable drift. Manoeuvre the wheel out from between the fork legs together with left- and right-hand spacers, noting their correct positions. Insert a wooden wedge between the brake pads to prevent their movement should the brake lever be accidentally operated.

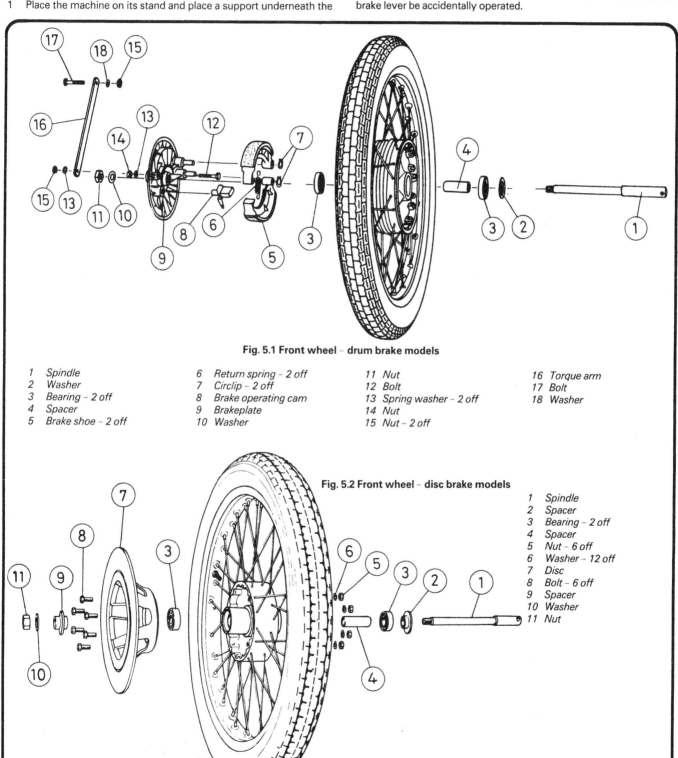

Fig. 5.1 Front wheel – drum brake models

1 Spindle	6 Return spring – 2 off	11 Nut	16 Torque arm
2 Washer	7 Circlip – 2 off	12 Bolt	17 Bolt
3 Bearing – 2 off	8 Brake operating cam	13 Spring washer – 2 off	18 Washer
4 Spacer	9 Brakeplate	14 Nut	
5 Brake shoe – 2 off	10 Washer	15 Nut – 2 off	

Fig. 5.2 Front wheel – disc brake models

1 Spindle
2 Spacer
3 Bearing – 2 off
4 Spacer
5 Nut – 6 off
6 Washer – 12 off
7 Disc
8 Bolt – 6 off
9 Spacer
10 Washer
11 Nut

3.2a On disc brake models do not omit wheel spacers before inserting spindle ...

3.2b ... then refit spindle washer and nut and tighten to the specified torque setting

3.4 Settle all components in position and tighten pinch bolt to specified torque setting

3   On models fitted with a front drum brake slacken the top torque arm mounting bolt and the nut which secures it to the brake plate and disconnect the arm from the brake plate. Remove the wheel spindle nut and washer, slacken its pinch bolt and withdraw the wheel spindle. If necessary the spindle can be tapped out of position using a hammer and suitable size drift. Partially remove the wheel until the brake plate assembly can be withdrawn from the wheel, then remove the wheel completely.
4   Refer to Routine maintenance for details of wheel examination.

3   Front wheel: refitting

1   Check that the wheel spindle is straight and free from corrosion before smearing a small amount of high melting-point grease along its shank to aid refitting.
2   On disc brake models remove the wooden wedge from the brake caliper and refit both the left and right-hand wheel spacers to the hub. Manoeuvre the wheel into position, ensuring that the disc is correctly positioned between the brake pads, insert the wheel spindle and refit its washer and nut, tightening it securely.
3   On drum brake models position the wheel between the fork legs and refit the brakeplate assembly to the wheel. Offer up the wheel and slide the large plain washer into position between the wheel and left-hand fork leg. Insert the wheel spindle and refit its washer and nut. Refit the torque arm to the brakeplate and tighten both its upper and lower mounting nuts and bolts. Apply the front brake to centralise the shoes then tighten the wheel spindle to the specified torque setting.
4   On all models remove the support from beneath the crankcase and push the machine off the stand. Apply the front brake and pump the forks up and down a few times to settle the disturbed components, then tighten the spindle pinch bolt to the specified torque setting. Check the operation of the front brake thoroughly before taking the machine on the road.

4   Rear wheel: removal

1   Place the machine on its centre stand and support it so that its rear wheel is clear of the ground and the machine will not tip backwards once the wheel is removed.
2   Disconnect the stop lamp switch wire (except 251 models) and remove the bolt which secures the torque arm to the brakeplate. Unscrew the brake rod adjusting nut, disengage the rod from the brake arm and remove the spring and washers for safekeeping.
3   Slacken the wheel spindle, withdraw it and remove the spacer from the left-hand side of the wheel. Pull the wheel to the left to disengage it from its cush drive assembly and position it so that the rear brakeplate assembly can be removed. The wheel can then be manoeuvred out of position noting that it may be necessary to tilt the machine slightly to the right in the process.

5   Rear wheel: refitting

1   Insert the wheel into the swinging arm, tilting the machine if necessary, and refit the brake plate assembly. Lift up the wheel, aligning its dogs with the holes in the cush drive assembly, and mesh the wheel with the cush drive. If this operation proves difficult the cush drive assembly can be held by selecting first gear. Check that the spindle is straight and free from corrosion and smear a small amount of high melting-point grease along its shank to aid refitting. Refit the spacer to the right-hand side of the hub and insert the spindle, tightening it finger-tight only at this stage.
2   Refit the washers and spring to the brake rod, fit the rod to the brake arm and locate the adjusting nut on the end of the rod. Refit the nut which secures the torque arm to the brakeplate. Apply the rear brake to centralise the brake shoes then tighten both the wheel spindle and torque arm bolt securely. Reconnect the stop lamp switch wire (except 251 model). Adjust the rear brake and check the operation of the stop lamp switch as described in Routine maintenance.

6   Wheel bearings: removal, examination and renovation

1   Remove the wheel from the machine as described in either Section 2 or 4 of this Chapter.
2   Position the wheel on a work surface with its hub well supported on wooden blocks so that enough clearance is left beneath the wheel to drive out the first bearing. Ensure the blocks are placed as close as

4.2 Do not forget to disconnect rear stop lamp switch wire (where fitted) before removing wheel

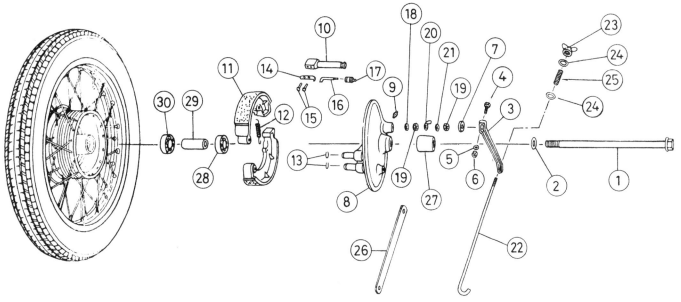

Fig. 5.3 Rear wheel

1 Spindle	9 Grease nipple	17 Insulating sleeve*	25 Spring
2 Washer	10 Brake operating cam	18 Insulating washer*	26 Torque arm (retaining
3 Brake operating lever	11 Brake shoe – 2 off	19 Nut – 2 off*	method differs between
4 Bolt	12 Return spring	20 Wire terminal*	models)
5 Spring washer	13 Circlip – 2 off	21 Washer*	27 Spacer
6 Nut	14 Stop lamp switch contact*	22 Brake rod	28 Bearing
7 O-ring	15 Screw – 2 off*	23 Adjusting nut	29 Spacer
8 Brakeplate	16 Switch contact screw*	24 Washer – 2 off	30 Bearing

*all models except 251

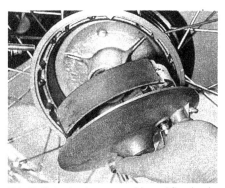

5.1a Refit brakeplate assembly and fit wheel to cush drive

5.1b Slide spacer into position and insert wheel spindle

5.2a Reconnect torque arm to brakeplate ...

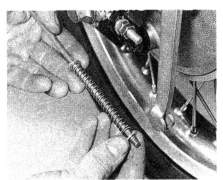

5.2b ... and refit the washers and spring to brake rod

5.2c Fit brake rod to brake arm and locate adjusting nut on end of the rod

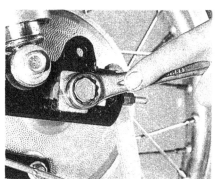

5.2d Apply rear brake to centralise shoes and tighten spindle securely

6.6a Wheel bearings should be fitted with their sealed surface facing outwards

6.6b Drift first bearing into position using a suitable size socket ...

6.6c ... then invert wheel, insert spacer and fit second bearing in a similar manner

possible to the bearing to avoid distorting the hub.

3   Insert a long drift down through the centre of the uppermost bearing and displace the central spacer to one side to allow the drift to rest on the inner race of the lower bearing. Using a hammer tap evenly around the inner race of the bearing until it is released from the hub. Invert the wheel, remove the central spacer and tap out the opposite bearing.

4   Remove all old grease from the hub and bearings by washing them with a high flash-point solvent. Check the bearings for freeplay between the inner and outer races, and for signs of roughness when rotated. If there is any doubt about the condition of either bearing, it should be renewed.

5   If the original bearings are to be refitted, then they should be repacked with a high melting-point grease before being fitted into the hub. The same also applies to new bearings. Ensure that the bearing recesses in the hub are clean and both bearings and recess mating surfaces are lightly greased to aid fitting. Check the condition of the hub recesses for signs of wear which may have been caused by the outer race of the bearing spinning. If evidence of this happening is found, and the bearing is a loose fit in the hub, then it is best to seek advice from an authorized MZ dealer or a competent motorcycle engineer. Alternatively a proprietary product such as Loctite Bearing Fit may be used to retain the bearing outer race; this will mean that the bearing housing must be carefully cleaned and degreased before the locking compound can be used.

6   With the hub and bearings thus prepared, fit the bearings and central spacer as follows. With the hub well supported by the wooden blocks, position the first of the bearings in the hub noting that both bearings must be fitted with their sealed surface outwards. Tap the bearing into place using a hammer and suitably sized socket or tube which bares only on the outer race of the bearing. With the first bearing in position, invert the wheel, insert the central spacer and pack the hub centre no more than ⅔ full with high melting-point grease. Tap the second bearing into position using the same procedure. Take great care to ensure that both bearings enter the hub squarely otherwise the bearing housings will be damaged.

7   Rear wheel sprocket, cush drive and speedometer drive: removal, examination and refitting

1   Remove the rear wheel as described in Section 4 of this Chapter.

2   Disengage the chain gaiters from the rear sprocket cover. Rotate the sprocket until the chain's connecting link comes into view, and disconnect it. Disengage the chain from the rear sprocket and temporarily rejoin the drive chain to prevent it from falling down either gaiter. Disconnect the speedometer cable from its drive gearbox and remove the sprocket retaining nut and washer. Remove the sprocket and cover assembly from the swinging arm and separate them noting the spacer fitted to the cover.

3   Refit the sprocket spindle nut to protect its threads and tap the spindle out of position. Remove the nut and withdraw the spindle. Release the speedometer drive gear from the sprocket by removing the circlip. If necessary the cush drive rubber can also be removed from the sprocket.

4   On 125 and 150 models the sprocket is fitted with a single bearing, retained by a circlip. This bearing can be drifted out once the circlip has been removed. However 250, 251 and 300 models the sprocket has two bearings. A bearing puller will almost certainly be necessary to remove the left-hand bearing although it may be possible to tap or lever it out once the sprocket has been warmed up gently. If this is attempted, take great care to avoid burning your hands on the hot sprocket. The right-hand bearing is retained by a circlip and can be drifted out of position once this has been removed. Check the bearing(s) for freeplay or signs of notchiness when rotated; renew if necessary.

5   Examine the rear sprocket for signs of wear and hooked or chipped teeth. If worn, the sprocket must be renewed along with the gearbox sprocket and drive chain. It is bad practice to renew just one sprocket on its own as running a new sprocket with a worn chain will quickly result in the new component becoming worn. Examine the cush drive rubber for signs of cracking or perishing and renew if worn.

6   Release the speedometer driven gear from the sprocket cover by unscrewing the countersunk screw from the outside of the cover. The

7.2a Disengage chain gaiters from sprocket cover ...

7.2b ... and disconnect speedometer drive cable

7.2c Remove sprocket cover retaining nut and remove cover from machine

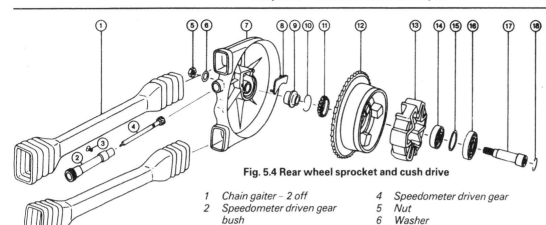

Fig. 5.4 Rear wheel sprocket and cush drive

1 Chain gaiter – 2 off
2 Speedometer driven gear bush
3 Screw
4 Speedometer driven gear
5 Nut
6 Washer
7 Sprocket housing

8 Cover
9 Spacer
10 Circlip
11 Speedometer drive gear
12 Rear sprocket
13 Cush drive rubber
14 Bearing
15 Circlip
16 Bearing (250, 251 and 300 only)
17 Spindle
18 Circlip

7.4 On 125 and 150 models, sprocket is fitted with single bearing, retained by circlip

7.5 Examine cush drive rubber for signs of wear and renew if necessary

7.6a Remove countersunk screw ...

7.6b ... and withdraw speedometer driven gear and bush

7.7a Ensure speedometer drive gear circlip fits into hole in sprocket

7.7b Refit sprocket spindle to the sprocket ...

7.7c ... followed by the spacer ...

7.7d ... then fit sprocket assembly to the cover

7.8 Tighten sprocket retaining nut securely once chain tension has been adjusted

bush and gear can then be withdrawn from the cover. Examine both the drive and driven gear for signs of damage, renewing as a pair if necessary. On reassembly smear the shaft of the speedometer driven gear with high melting-point grease and insert it into the bush. Fit the bush and gear to the cover and secure them with the countersunk screw and the plastic cover.

7   The sprocket assembly is reassembled by a reverse of the removal procedure. Pack the bearing(s) with high melting-point grease and apply a smear of grease to the shank of the spindle to aid refitting. Tap the bearing into the sprocket, using a hammer and a suitable socket which bears only on the outer race of the bearing, until its retaining circlip can be refitted. On 250, 251 and 300 models drift the second bearing into position in a similar manner. On all models insert the spindle into the bearing(s) and tap it fully into place. Refit the speedometer drive gear to the outside of the sprocket and retain it with its circlip. Refit the sprocket cover and spacer and fit the cush drive rubber to the sprocket assembly.

8   Fit the sprocket assembly to the swinging arm and refit its retaining nut and washer finger-tight only at this stage. Refit the chain to the rear sprocket, then refit its connecting link ensuring that the spring clip is fitted with its closed end facing the normal direction of chain travel. Refit the gaiters to the sprocket cover and reconnect the speedometer cable. Refit the rear wheel as described in Section 5 and adjust the chain and wheel alignment as described in Routine maintenance.

8 Brake disc: examination and renovation

1   The brake disc can be checked for wear and warpage without removing the wheel from the machine. Using a micrometer, measure the thickness of the disc at several points around the disc. If any measurement taken is less than the specified service limit, or if the difference between any two measurements is greater than the maximum difference specified, the disc must be renewed. Note all brake disc measurements should be taken approximately 10 mm (0.4 in) in from the outer edge of the disc. Check the warpage (runout) of the disc by setting a suitable pointer close to the outer periphery of the disc and spinning the front wheel slowly. If disc warpage exceeds the specified limit, the disc must be renewed. A warped disc, apart from reducing the braking efficiency, is likely to cause juddering during braking and will cause the brake to bind when not in use.

2   The brake disc should also be checked for bad scoring on its contact area. If found, the disc should be renewed or, if possible, repaired by skimming. Such repairs should only be entrusted to a reputable engineering firm. A local motorcycle firm may be able to assist in having the work carried out.

3   To remove the disc, first remove the front wheel as described in Section 2. Slacken and remove the six disc mounting nuts and bolts and lift the disc off the wheel. On refitting, renew the six self-locking nuts and tighten them evenly and securely. Refit the wheel as described in Section 3 of this Chapter.

8.3 Brake disc is retained by six bolts

9 Master cylinder: examination and renovation

Note: *Take great care not to spill hydraulic fluid on any painted or plastic cycle parts; it is a very effective paint stripper.*

1   If regular checks of the braking system reveal the presence of a leak, or if a loss of brake pressure is encountered at anytime, the handlebar-mounted master cylinder assembly must be removed and dismantled for inspection.

2   Disconnect the stoplamp switch wires from the switch on the underside of the assembly. The switch itself need not be disturbed unless the master cylinder is to be renewed. Place a clean container beneath the caliper unit and run a clear plastic tube from the caliper bleed nipple to the container. Unscrew the bleed nipple by one turn and pump the brake lever repeatedly until no more fluid can be seen issuing from the bleed nipple.

3   Position a clean rag below the point where the brake hose joins the master cylinder, to prevent brake fluid dripping onto the machine, and disconnect the brake hose. Once the excess fluid has drained from the master cylinder place the end of the hose in a polythene bag and tie it to the handlebars.

4   Remove the circlip from the brake lever pivot pin, withdraw the pin and remove the brake lever. Unscrew the reservoir cap and remove the ventilation ring and the diaphragm. Release the two bolts from the master cylinder mounting clamp and remove the master cylinder assembly from the handlebars.

5   To remove the piston assembly from the cylinder bore it is likely that the special MZ service tool will be required. It is therefore recommended that the master cylinder be taken to an authorized MZ dealer who will have access to the tool. However, if lucky, it may be possible to remove the piston as follows. Using a small flat-bladed screwdriver, carefully prise out the circlip which retains the piston seal in the cylinder. Spray the seal with a penetrating fluid, such as WD40, and allow it to soak for a few minutes. Then using a pair of pointed nose pliers, grip the end of the piston firmly and try to pull it out of the master cylinder body. At first this will take quite a large amount of force to dislodge the seal but after that the piston should withdraw fairly easily. Once the piston is removed, withdraw the return spring from the caliper bore.

6   Thoroughly clean all the master cylinder components and inspect them as follows. Inspect the master cylinder body for signs of stress

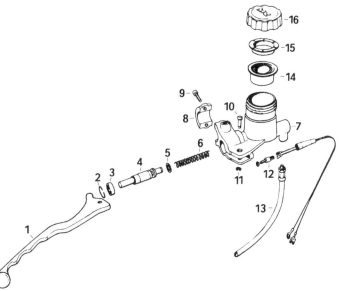

Fig. 5.5 Front brake master cylinder

1	Brake lever	9	Bolt – 2 off
2	Circlip	10	Brake lever pivot
3	Seal	11	Circlip
4	Primary piston	12	Brake light switch
5	O-ring	13	Brake hose
6	Return spring	14	Diaphragm
7	Master cylinder	15	Ventilation ring
8	Mounting clamp	16	Reservoir cap

fractures around the brake lever pivot lugs and the mounting point. If cracks are detected the unit body must be renewed. Clean the piston components and master cylinder bore with fresh brake fluid and examine each component for signs of pitting or scoring, renewing as necessary. The piston return spring can only be checked in comparison with a new component and should be renewed if in any doubt about its condition. Renew the piston O-ring and caliper seal regardless of their apparent condition.

7   During reassembly cleanliness is essential to prevent dirt from entering the hydraulic system. Fit the return spring in the cylinder bore and fit the new O-ring to the piston. Liberally lubricate the surfaces of the piston and cylinder bore with clean hydraulic fluid and insert the piston. Lubricate the cylinder seal with clean hydraulic fluid and fit it to the cylinder body. The seal can be pressed or tapped into position using a long reach socket or piece of tubing which bears only on the hard outer edge of the seal. Once the seal is correctly seated refit the circlip, ensuring that it is correctly fitted in its groove. Check that the piston moves smoothly and returns quickly.

8   Refit the master cylinder to the handlebars, positioning it so that the reservoir is horizontal during normal use, and tighten its mounting bolts securely. Refit the brake lever, securing the pivot pin with its circlip, and reconnect the brake hose. Reconnect the stop lamp switch wire.

9   Fill the reservoir with new hydraulic fluid and bleed the brake as described in Section 12. On completion of the bleeding process, check the master cylinder for leaks whilst applying the front brake. If all appears to be well, thoroughly check the operation of the front brake before taking the machine on the road. During the run, frequently use the brake and recheck for leaks afterwards.

10   Brake hose: examination and renovation

When the brake assembly is being overhauled, or at anytime during a routine maintenance or cleaning procedure, check the brake hose for signs of leakage, damage, deterioration or scuffing against any cycle components. If any such damage is discovered, the hose must be renewed immediately. The union connections at each end of the hose must also be in a good condition, with no stripped or damaged threads. Examine all sealing washers and renew any which are damaged. Do not overtighten the hose unions as they are easily damaged and could shear.

11   Brake caliper: examination and renovation

Note: *Take great care not to spill hydraulic fluid on any painted or plastic cycle parts; it is a very effective paint stripper.*

1   If the regular checks described in Routine maintenance reveal the presence of fluid leakage or of any wear or damage which will impair the caliper's efficiency, the unit must be removed and dismantled for inspection as follows.

2   Attach a length of clear plastic tubing to the caliper bleed nipple, placing the lower end of the tube in a suitable container. Open the bleed nipple by one full turn and gently pump the front brake lever to drain as much fluid as possible from the hydraulic system. When no more fluid can be seen issuing from the bleed nipple, tighten the nipple and remove the plastic hose from the caliper. Slacken the brake hose union nut and disconnect the hose from the caliper. Place the end of the hose in a clean polythene bag and tie it to the fork leg. Immediately wash off any brake fluid which has been spilt. Remove the caliper mounting bolts, slide the caliper off the disc and remove the brake pads as described in Routine maintenance.

3   Slacken and remove the two bolts which secure the two halves of the caliper and separate them noting the small O-ring fitted to the internal passage. Dismantle each half of the caliper separately to avoid interchanging components.

4   The pistons can be removed using a jet of compressed air. Wrap a clean rag around the caliper half and apply the air into the internal passage of the caliper, noting that on the outer half it will be necessary to block the caliper union. Be careful not to use too high an air pressure to expel the pistons or the pistons may become damaged. Under no circumstances should the pistons be levered out of their bores. If the above method fails to dislodge the pistons, no further time should be wasted and the complete caliper assembly must be renewed. Even if a method is devised to remove the pistons, it is likely that the caliper bore and piston are so badly corroded that they will have to be renewed anyway.

5   When each piston has been removed, it should be placed in a clean container and the seals should be picked out of the caliper bores. Clean the caliper bore and piston and inspect them for signs scoring, pitting or corrosion. If any of these defects are evident it is unlikely that a good fluid seal can be maintained and for this reason the damaged components should be renewed. Renew the piston seals and internal O-ring as a matter of course.

6   On reassembly, cleanliness is essential to prevent the entry of dirt into the hydraulic system. Soak the new seals in clean hydraulic fluid and press them into position in the caliper bore until they are properly seated. Smear a liberal quantity of clean brake fluid over the surface of the pistons and caliper bores then carefully insert the pistons with a twisting motion, as if screwing them in, to avoid damaging the seals. Once each piston is fitted in its original caliper half, wipe the mating surfaces of each half clean and fit a new O-ring to the internal passage. Join the caliper halves, refit the two bolts and washers and tighten them securely. Wipe off any surplus brake fluid from the caliper and refit the brake pads as described in Routine maintenance.

7   Refit the caliper assembly to the machine and reconnect the brake hose. Fill the master cylinder reservoir with new hydraulic fluid and bleed the system as described in Section 12. On completion of bleeding, carry out a check for fluid leakage whilst applying the front brake. If all is well, thoroughly check the operation of the front brake before taking the machine on the road. During the run, use the brake frequently and recheck for leaks afterwards.

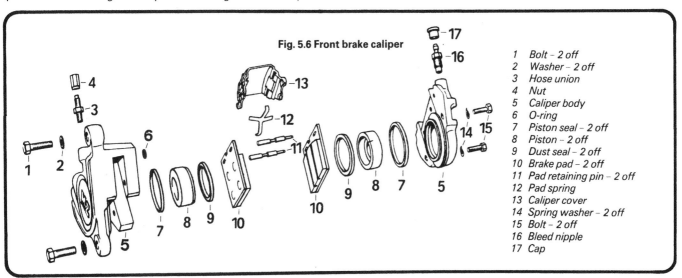

Fig. 5.6 Front brake caliper

1   Bolt – 2 off
2   Washer – 2 off
3   Hose union
4   Nut
5   Caliper body
6   O-ring
7   Piston seal – 2 off
8   Piston – 2 off
9   Dust seal – 2 off
10 Brake pad – 2 off
11 Pad retaining pin – 2 off
12 Pad spring
13 Caliper cover
14 Spring washer – 2 off
15 Bolt – 2 off
16 Bleed nipple
17 Cap

12   Bleeding the hydraulic brake system

Note: *Take great care not to spill hydraulic fluid on any painted or plastic cycle parts; it is a very effective paint stripper.*

1   If brake action becomes spongy, or if any part of the hydraulic system is dismantled (such as when a hose is renewed) it is necessary to bleed the system in order to remove all traces of air. The procedure for bleeding the hydraulic system is best carried out by two people.
2   Check the fluid level in the reservoir and top up with new fluid of the specified type if required. Keep the reservoir at least half full during the bleeding procedure; if the level is allowed to fall too far air will enter the system requiring that the procedure be started again from scratch. Refit the reservoir cap to prevent the ingress of dust or the ejection of a spout of fluid.
3   Remove the dust cap from the caliper bleed nipple and clean the area with a rag. Place a clean glass jar below the caliper and connect a pipe from the bleed nipple to the jar. A clear plastic pipe should be used so that air bubbles can be more easily seen. Pour enough clean hydraulic fluid in the glass jar so that the pipe end is immersed below the fluid surface; ensure that the pipe end remains submerged (to prevent air returning to the system whenever the pressure is released) throughout the operation.
4   If parts of the system have been renewed, and thus the system must be filled, open the bleed nipple about one turn and pump the brake lever until fluid starts it issue from the clear pipe. Tighten the bleed nipple and then continue the normal bleeding operation as described in the following paragraphs. Keep a close check on the reservoir level whilst the system is being filled.
5   Operate the brake lever as far as it will go and hold it in this position against the fluid pressure. If spongy brake operation has occurred it may be necessary to pump the brake lever rapidly a number of times until pressure is built up. With pressure applied, loosen the bleed nipple about half a turn. Tighten the nipple as soon as the lever has reached its full travel and then release the lever. Repeat this operation until no more air bubbles are expelled with the fluid into the glass jar. When this condition is reached the air bleeding operation should be complete, resulting in a firm feel to the brake lever. If sponginess is still evident continue the bleeding operation; it may be that an air bubble trapped at the top of the system has yet to work down through the caliper.
6   When all traces of air have been removed from the system, top up the reservoir and refit the diaphragm, ventilation ring and cap. Check the entire system for leaks, and check also that the brake system in general is functioning efficiently before using the machine on the road.
7   Brake fluid drained from the system will almost certainly be contaminated, either by foreign matter or more commonly by the absorption of water from the air. All hydraulic fluids are to some degree hygroscopic, that is, they are capable of drawing water from the atmosphere, and thereby degrading their specifications. In view of this, and the relative cheapness of the fluid, old fluid should always be discarded.

13   Drum brakes: dismantling, examination and reassembly

Front brake
1   Remove the front wheel as described in Section 2 of this Chapter.

2   Mark the brake shoes in some way so that they can be refitted in their original positions, remove their retaining circlips and lift the shoes off the brakeplate. Pull the operating arm/cam out of the brake plate, disconnect it from the brake cable and remove the brakeplate and operating arm/cam from the machine.
3   Examine the condition of the brake linings. If oil or grease from the wheel bearings has badly contaminated the linings, the brake shoes must be renewed. There is no satisfactory way of degreasing the friction material. Any surface dirt on the linings can be removed with a stiff bristled brush. High spots on the linings should be eased down using emery cloth. MZ do not specify a minimum friction material thickness for these models. However, it is recommended that if either shoe's friction material has worn to 1.5 mm (0.06 in) or less at any point along its length, a figure given for earlier MZ models, the shoes should be renewed as a set. If the existing shoes are to be re-used, roughen up the friction material surface sufficiently to break the glaze which will have formed in use. Emery cloth is ideal for this purpose but take care not to inhale the asbestos dust which comes off the linings.
4   Remove all traces of dust from the brake drum, preferably using a brass wire brush. Take care not to inhale any of the dust as it is of an asbestos nature and consequently harmful. Wipe the drum with a rag soaked in solvent to remove any traces of oil or grease. Inspect the surface of the drum for signs of scoring. If deep scoring is evident, the drum must either be skimmed on a lathe or renewed. Whilst MZ do not specify a maximum diameter for the drum, it should be borne in mind that excessive skimming will change the radius of the drum in relation to the shoes, therefore reducing the friction area until extensive bedding in has taken place. Also full adjustment of the brake shoes may not be possible. If in doubt about this point, the advice of one of the specialist engineering firms who undertake this work should be sought.
5   Examine the brake operating arm/cam and brakeplate for signs of wear or damage, renewing if necessary. The brake shoe return spring can only be tested in comparison with a new component and should be renewed if there is the slightest doubt about its condition.
6   Note that it is false economy to try to cut corners with brake components; the whole safety of both rider and machine being dependent on their condition.
7   The brakeplate assembly is reassembled by a reverse of the removal sequence. Insert the brake cable through the plate and refit the operating arm/cam to the cable. Apply a small amount of high melting-point grease to the operating arm/cam pivot and refit it to the brakeplate. Apply a smear of high melting-point grease to the brake shoe pivot points and operating cam and fit the return spring to the brake shoes. Fit the shoes to the brakeplate using the marks made on dismantling, and secure them with the circlips. Operate the front brake lever a few times to make sure that the brake shoes are free to move smoothly and return quickly.

Rear brake
8   Remove the rear wheel from the machine as described in Section 4 of this Chapter and remove the brakeplate assembly.
9   Remove the two circlips which retain the brake shoes and lift them off the brakeplate. Mark the brake shoes in some way so they can be refitted in their original positions. Mark the brake operating arm and shaft, remove the operating arm pinch bolt and pull the arm off the shaft. Remove the camshaft from the brakeplate.
10   Examine all rear brake components as described in paragraphs 3 – 6 above, renewing as necessary.

13.4 Brake drum must be clean and free from scoring

13.9a Remove circlips ...

13.9b ... and withdraw brake shoes

13.11 Check condition of stoplamp switch contacts (where fitted) and renew if necessary

11   On all models except the 251, the rear stoplamp switch is fitted in the brakeplate. The terminal fitted to the brakeplate should come in contact with the small plate on the camshaft retained by two screws. If either the terminal or plate are worn or corroded they must be renewed. If both are in a good condition, clean the terminal and plate to remove all traces of grease and dust and reassemble the brakeplate as follows.
12   Apply a smear of high melting-point grease to the camshaft and insert it into the brakeplate. Refit the operating arm, using the marks made on dismantling to position it correctly, and tighten its pinch bolt securely. Apply a smear of high melting-point grease the brake shoe pivot points and fit the return spring to the brake shoes. Fit the shoes to the brakeplate and secure them with their circlips. Move the operating arm and check that the shoes move smoothly and return quickly.

Front and rear brake
13   Refit the wheel as described in Section 2 or 4 of this Chapter, and adjust the brake and stoplamp switch as described in Routine maintenance. Thoroughly check the operation of the brake before taking the machine on the road, remembering that if new brake shoes have been fitted the shoes will need to bed in before normal braking performance returns. To a lesser extent, the same applies to shoes which have been cleaned and re-used.

---

14   Tyres: removal, puncture repair and refitting

---

1   To remove the tyre from either wheel first detach the wheel from the machine. Deflate the inner tube by removing the valve core, and when the tyre is fully deflated, push the bead away from the wheel rim on both sides so that the bead enters the centre well of the rim. Remove the locking ring and push the tyre valve into the tyre itself.
2   Insert a tyre lever close to the valve and lever the edge of the tyre over the outside of the rim. Very little force should be necessary; if resistance is encountered it is probably due to the fact that the tyre beads have not entered the well of the rim all the way round. If aluminium rims are fitted, damage to the soft alloy by tyre levers can be prevented by the use of plastic rim protectors. These can be fabricated from short lengths (4 - 6 inches) of thick-walled nylon petrol pipe which has been split down one side using a sharp knife.
3   Once the tyre has been over the wheel rim, it is easy to work around the rim, so that the tyre is completely free from one side. At this stage the inner tube can be removed.
4   Now working from the other side of the wheel, ease the other edge of the tyre over the outside of the wheel rim that is furthest away. Continue to work around the rim until the tyre is completely free from the rim.
5   If a puncture has necessitated the removal of the tyre, reinflate the inner tube and immerse it in a bowl of water to trace the source of the leak. Mark the position of the leak and deflate the tube. Dry the tube and

clean the area around the puncture with a rag moistened in petrol. Roughen the area where the patch will be applied to obtain good adhesion and apply rubber solution and allow this to dry before removing the backing from the patch, and applying the patch to the surface.
6   It is best to use a patch of the self vulcanizing type, which will form a permanent repair. Note that it may be necessary to remove a protective covering from the top surface of the patch after it has sealed into position. Inner tubes made from a special synthetic rubber may require a special type of patch and adhesive, if a satisfactory bond is to be achieved.
7   If the inner tube has been patched on a number of past occasions, or if there is a tear or large hole, it is preferable to discard it and fit a new tube. Sudden deflation may cause an accident, particularly if it occurs with the rear wheel.
8   Before refitting the tyre, check the inside to make sure that the article that caused the puncture is not still trapped inside it. Check the outside of the tyre, particularly the tread area to make sure that nothing is trapped that may cause a further puncture.
9   Never fit a tyre that has a damaged tread or sidewalls. Apart from the legal aspects there is a very great risk of a blowout, which can have serious consequences on a two wheeled vehicle.
10   Ensure that the rim tape is in position. If this precaution is overlooked there is a good chance of the spoke nipple ends chafing the inner tube and causing a crop of punctures.
11   Inflate the inner tube for it just to assume a circular shape, but only to that amount, and then push the tube into the tyre so that it is enclosed completely. Lay the tyre on the wheel at an angle, taking note of any direction of rotation arrows on the sidewall, and insert the valve through the rim tape and the hole in the wheel rim. Attach the locking ring on the first few threads, sufficient to hold the valve captive in its correct position.
12   Tyre refitting is aided by dusting the sidewalls, particularly in the vicinity of the beads, with a liberal coating of french chalk.
13   Starting at the point furthest from the valve, push the tyre bead over the edge of the wheel rim until it is located in the centre well. Continue to work around the tyre in this fashion until the whole of one side of the tyre is on the rim. It may be necessary to use a tyre lever during the final stages.
14   Make sure there is no pull on the tyre valve and again commencing with the area furthest from the valve, ease the other bead of the tyre over the edge of the rim. Finish with the area close to the valve, pushing the valve up into the tyre until the locking ring touches the rim. This will ensure that the inner tube is not trapped when the last section of bead is edged over the rim with a tyre lever.
15   Check that the inner tube is not trapped at any point. Reinflate the inner tube, and check that the tyre is seating correctly around the wheel rim. There should be a thin rib moulded around the tyre on both sides, which should be an equal distance from the wheel rim at all points. If the tyre is unevenly located on the rim, try bouncing the wheel when the tyre is at the recommended pressure. It is probable that one of the tyre beads has not pulled clear of the centre well.
16   Always run the tyres at the recommended pressures and never under or over inflate. The correct pressures are given in Routine maintenance; if non-standard tyres are fitted check with the tyre manufacturer or supplier for recommended pressures. Finally refit the valve dust cap.

---

15   Tyre valves: examination and renewal

---

1   Valve cores and caps seldom give trouble, but do not last indefinitely. Dirt under the seating will cause a puzzling 'slow-puncture'. Check that they are not leaking by applying spittle to the end of the valve and watching for air bubbles.
2   The valve core screws into the valve body and can be removed with a small slotted tool which is normally incorporated in plunger type pressure gauges. Some valve dust caps incorporate a key for removing valve cores. Occasionally, an elusive slow puncture can be traced to a leaking valve core, and this should be checked before a genuine puncture is suspected. Check that the valve retaining nut is securely fastened against the rim.
3   The valve cap is a significant part of the tyre valve assembly. Not only does it prevent the ingress of road dirt, but also acts as a secondary seal which will reduce the risk of sudden deflation if a valve core should fail.

16   Wheel balancing

1   It is customary to balance the wheels complete with the tyre and inner tube. The out of balance forces which exist are eliminated and the handling of the machine is improved in consequence. A wheel which is badly out of balance produces through the steering a most unpleasant hammering effect at high speeds.

2   Some tyres have a balance mark on the sidewall, usually in the form of a coloured spot. This mark must be in line with the tyre valve when the tyre is fitted to the rim. Even then the wheel may require the addition of balance weights to offset the weight of the tyre valve itself.

3   If the wheel is raised clear of the ground and is spun, it will probably come to rest with the tyre valve or the heaviest part downward and will always come to rest in the same position. Balance weights must be added to a point diametrically opposite this heavy spot until the wheel comes to rest in *any* position after it has been spun. These weights are available from authorized MZ service agents and are fitted to the spokes. Note on the rear wheel it is recommended that the wheel is removed from the machine to be balanced, or alternately the drive chain must be disconnected and removed from the rear sprocket. Otherwise the friction of the chain will influence where the rear wheel stops.

Tyre changing sequence - tubed tyres

 Deflate tyre. After pushing tyre beads away from rim flanges push tyre bead into well of rim at point opposite valve. Insert tyre lever adjacent to valve and work bead over edge of rim.

Use two levers to work bead over edge of rim. Note use of rim protectors

 Remove inner tube from tyre

When first bead is clear, remove tyre as shown

 When fitting, partially inflate inner tube and insert in tyre

Work first bead over rim and feed valve through hole in rim. Partially screw on retaining nut to hold valve in place.

 Check that inner tube is positioned correctly and work second bead over rim using tyre levers. Start at a point opposite valve.

Work final area of bead over rim whilst pushing valve inwards to ensure that inner tube is not trapped

Chapter 6 Electrical system

Refer to Chapter 7 for information relating to 1991-on models

Contents

Specifications

Battery

Capacity:	
125 and 150 models	12V 5.5AH
250, 251 and 300 models	12V 5.5AH or 12V 9AH
Earth..	Negative
Specific gravity...	1.280

Generator

Type...	Three-phase
Maximum output..	14 volts, 15 amps
Carbon brushes:	
Standard length	16 mm (0.63 in)
Service limit..	9 mm (0.35 in)
Stator coil winding resistance	Approximately 0.32 ohms
Rotor coil winding resistance	4.2 ± 0.3 ohms
Slip rings:	
Minimum diameter	31 mm (1.22 in)
Maximum runout	0.05 mm (0.002 in)

Fuses

Main	16A x 2
Turn signal...	4A
Charging system..	2AT

Bulbs

Headlamp	12V 45/40W or 12V 55/60W H4
Parking lamp	12V 4W
Stop lamp*	12V 21W
Tail lamp*	12V 5W
Turn signal lamp	12V 21W
Instrument and warning lamps	12V 2W

* Note: 251 models are fitted with a twin filament 12V 5/21W stop/tail lamp bulb instead of two separate bulbs.

1 General description

Power for the complete electrical system is provided by a three-phase alternator mounted on the crankshaft right-hand end. Its output is rectified and controlled by separate rectifier and regulator units before being passed to the battery and thence to the main electrical system.

2 Electrical system: general information and preliminary checks

1   In the event of an electrical system fault, always check the physical condition of the wiring and connectors before attempting any of the test procedures described here and in subsequent sections. Look for chafed, trapped or broken electrical leads and repair or renew as necessary.

Leads which have broken internally are not easily spotted, but may be checked using a multimeter or a simple battery and bulb circuit as a continuity tester. This arrangement is shown in the accompanying illustration. The various multi-pin connectors are generally trouble-free but may corrode if exposed to water. Clean them carefully, scraping off any surface deposits, and pack with silicone grease during assembly to avoid recurrent problems. The same technique can be applied to the handlebar switches.

2   The wiring harness is colour-coded and will correspond with the wiring diagrams at the end of this manual. Where socket connections are used, they are designed so that reconnection can be made only in the correct position.

3   Visual inspection will usually show whether there are any breaks or frayed outer coverings which will give rise to short circuits. Occasionally a wire may become trapped between two components, breaking the inner core but leaving the more resilient outer cover intact. This can give rise to mysterious intermittent or total circuit failure. Another source of trouble may be the snap connectors or sockets, where the connector has not been pushed fully home in the outer housing, or where corrosion has occurred.

4   Intermittent short circuits can often be traced to a chafed wire that passes through or is close to a metal component such as a frame member. Avoid tight bends in the lead or situations where a lead can become trapped between casings.

5   A sound, fully charged battery, is essential to the normal operation of the system. There is no point in attempting to locate a fault if the battery is partly discharged or worn out. Check battery condition and recharge or renew the battery before proceeding further.

6   Many of the test procedures described in this chapter require voltages or resistances to be checked. This necessitates the use of some form of test equipment such as a simple and inexpensive multimeter of the type sold by electronics or motor accessory shops.

7   If you doubt your ability to check the electrical system, entrust the work to an authorized MZ dealer. In any event have your findings double-checked before consigning expensive components to the scrap bin.

3   Battery: examination and maintenance

1   Details of the regular checks needed to maintain the battery in good condition are given in Routine maintenance, together with instructions on removal and refitting and general battery care. Batteries can be dangerous if mishandled; read the Safety first! section at the front of this manual before starting work, and always wear overalls or old clothing in case of accidental acid spillage. If acid is ever allowed to splash into your eyes or onto your skin, flush it away with copious quantities of fresh water and seek medical advice immediately.

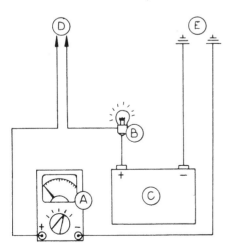

Fig. 6.1 Simple circuit testing equipment (Sec 2)

A   Multimeter   D   Positive probe
B   Bulb   E   Negative probe
C   Battery

2   When new, the battery is filled with an electrolyte of dilute sulphuric acid having a specific gravity of 1.280 at 20°C (68°F). Subsequent evaporation, which occurs in normal use, can be compensated for by topping up with distilled or demineralised water only. Never use tap water as a substitute and do not add fresh electrolyte unless spillage has occurred.

3   The state of charge of a battery can be checked using an hydrometer.

4   The normal charge rate for a battery is $\frac{1}{10}$ of its rated capacity, thus for a 5 ampere hour unit charging should take place at 0.5 amp. Exceeding this figure can cause the battery to overheat, buckling the plates and rendering it useless. Few owners will have access to an expensive current controlled charger, so if a normal domestic charger is used check that after a possible initial peak, the charge rate falls to a safe level. *If the battery becomes hot during charging stop; further charging will cause damage.* Note that the cell caps should be loosened and the vents unobstructed during charging to avoid a build-up of pressure and risk of explosion.

5   After charging top up with distilled water as required, then check the specific gravity and battery voltage. Specific gravity should be above 1.270 and a sound, fully charged battery should produce 15 - 16 volts. If the recharged battery discharges rapidly if left disconnected it is likely that an internal short caused by physical damage or sulphation has occurred. A new battery will be required. A sound item will tend to lose its charge at about 1% per day.

4   Charging system: checking the output

1   If a battery, which is known to be in a good condition, fails to hold its charge when fitted to the machine the generator output should be checked as follows. Note that the machine must be fitted with a fully charged battery if the test is to prove conclusive.

2   Remove the right-hand sidepanel and connect a multimeter, set to the 20V dc scale, across the battery terminals (meter positive probe to battery positive terminal (+) and meter negative probe to battery negative terminal (-). Start the engine and slowly increase the engine speed whilst noting the readings obtained. Stop the engine. If the charging system is in a good condition a reading of approximately 13 - 14.5 volts should be obtained, which should remain fairly constant at all engine speeds. If the reading obtained is not within this range, the charging system components should be tested in the order given below. Should the generator output prove to be excessive note that the most likely cause of the problem is the regulator which should be examined first as described in Section 7.

(a)   Generator carbon brushes, see Section 5.
(b)   Charging system wiring and connectors, see Section 2.
(c)   Generator stator coil, see Section 5.
(d)   Generator rotor coil, see Section 5.
(e)   Rectifier unit, see Section 6.
(f)   Regulator unit, see Section 7.

5   Generator: overhaul

1   Alternator components are removed and refitted as described in Sections 7 and of 35 Chapter 1. Examine the components as described under the relevant sub-heading.

Carbon brushes
2   Press the carbon brushes into the holder and remove the retaining clip from holder. Then slowly release the brushes and withdraw them from the brushholder along with their return springs.

3   Measure the length of each carbon brush using a vernier caliper. If either brush has worn beyond, or close to, its service limit, both brushes should be renewed as a set. MZ specify that the brush return springs should exert a force of 1.4 - 3.2 Nm. This can be checked if the necessary equipment is available. If not, the springs should be renewed if there is any doubt about their condition.

4   On reassembly, refit the return springs and brushes to the holder and secure them with the retaining clip. Ensure the brushes move freely in the holder and are pushed firmly against the slip rings of the rotor once refitted.

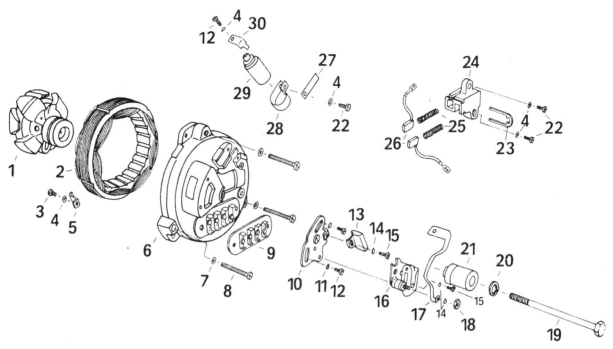

Fig. 6.2 Generator

1	Rotor	9	Terminal plate
2	Stator coil	10	Contact breaker mounting
3	Screw – 3 off		plate
4	Spring washer – 7 off	11	Spring washer – 3 off
5	Clamp – 3 off	12	Screw – 2 off
6	Stator coil housing	13	Cam wiper felt
7	Spring washer – 3 off	14	Spring washer
8	Screw – 3 off	15	Screw – 2 off

16	Contact breaker	24	Brush holder
17	Connecting wire	25	Spring – 2 off
18	Nut	26	Brush – 2 off
19	Rotor retaining bolt	27	Clip
20	Spring washer	28	Capacitor mounting clamp
21	Cam	29	Capacitor
22	Screw – 3 off	30	Capacitor terminal
23	Brush retaining clip		

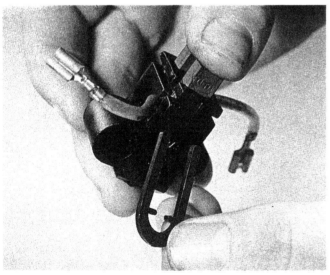

5.2 Remove retaining clip to release brushes from holder

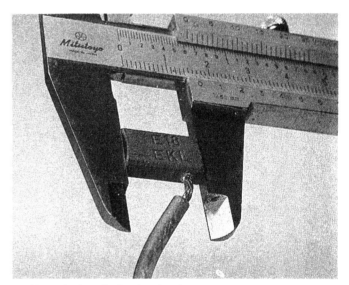

5.3 Measuring length of carbon brushes

Stator windings

5   Using a multimeter set to the ohm x 1 scale, check for stator windings for short or open circuits by in turn measuring the resistance between the U, V, and W terminals of the stator housing, so that a total of three different measurements are taken. All three of the readings obtained should conform to that given in the Specifications. If the resistance between any two of the wires greatly exceeds this figure, or a reading of 0 ohms (indicating a short circuit) is obtained, the stator coil

windings can be considered faulty and must be removed and renewed.
6   To remove the stator coil windings from their housing, unsolder the windings from the U, V and W terminals, and slacken and remove its three retaining screws and clamps. On refitting, solder the three wires to the U,V and W terminals of the housing, and ensure that the groove in the stator windings is correctly aligned with the groove in the housing. Refit the three clamps which retain the windings and tighten their retaining screws securely.

Rotor windings

7 Using a multimeter set to the ohms x 10 scale, check the rotor coil windings by measuring the resistance between the two slip rings. *Do not press the meter probes against the slip rings with any force as they are easily damaged.* If the reading obtained differs considerably from that given in the Specifications, the rotor windings can be considered faulty and must be renewed.

8 Examine the slip rings for signs of pitting or scoring. If necessary, tidy up the slip rings using a piece of fine emery cloth. Using a pair of vernier calipers, measure the diameter of each slip ring. If either slip ring has worn below the service limit given in the Specifications, the rotor coil should be renewed.

6 Rectifier: testing

1 The rectifier is located underneath the seat, next to the ignition HT coil. To test the rectifier it is first necessary to remove it from the machine by disconnecting its electrical connections and removing its mounting screws.

2 Make a note of how the wires fitted to the U, V and W terminals are arranged, then unsolder them from the terminals. Slacken and remove the four screws which secure the three sections of the rectifier together and separate the rectifier components taking care not to lose the

6.1 Rectifier is located under seat and is retained by two screws

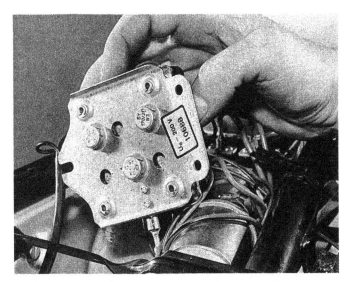

6.4 Positive and negative diodes can be renewed individually

spacers which are fitted between the plates.

3 To test the rectifier diodes a 12 volt battery and bulb arrangement is needed. Note that the test bulb should be no more than 15 W for the test to prove accurate. To test the negative (bottom) diode plate, connect the battery negative (–) terminal to the D- terminal of the plate and then connect the battery positive (+) terminal to each of the diodes in turn. With the battery and bulb connected in this way the bulb should not light up at all. Then reverse the battery connections, battery positive (+) to the D- terminal and battery negative (–) to each of the diodes in turn noting the results. With the battery and bulb connected in this way the bulb should light as the negative lead contacts each diode. The positive (centre) diode plate is tested in the same way but the results should be reversed, ie the bulb should light when the battery negative (–) lead is connected to the D + terminal and the battery (+) positive lead to each of the diodes and not when the battery leads are reversed. If the bulb does not perform as described above when connected to any diode, that diode is faulty and must be renewed.

4 Both positive and negative diodes can be renewed individually. However, to remove the old diodes and fit the new items a hydraulic press and mandrels are needed. Because of this, and the fact that the components are so delicate and are easily damaged, it is recommended that the rectifier components should be taken to an authorized MZ service agent, who will have the relevant equipment to remove and fit the diodes, should diode renewal be necessary.

5 The exciter diodes fitted to the insulating (top) plate can also be tested as described above and should give the same results as the positive diode plate, ie the bulb should light when the battery negative (–) lead is connected to the 61 terminal and the battery positive (+) lead to each of the diodes in turn and not when the battery leads are reversed. If the above test proves any of the diodes to be faulty they must be renewed. Old diodes should be unsoldered and the new items soldered in place.

6 Once all the diodes are known to be serviceable, reassemble the rectifier components and position the spacers between the plates. Refit the four screws which secure the rectifier components together and tighten them securely. Resolder the connections to the U, V and W terminals of the insulating plate, using the notes made on dismantling, and coat all soldered connections with a smear of silicone grease. Refit the rectifier unit to the machine, tightening its mounting bolts securely, and reconnect all its electrical connections.

7 Regulator: overhaul and testing

1 The regulator is located beneath the seat and should require very

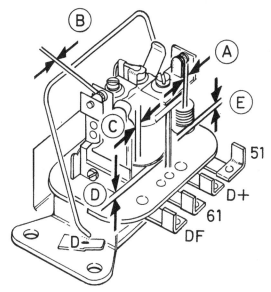

Fig. 6.3 Regulator mechanical settings (Sec 7)

A = 1.4 – 1.5 mm   D = 0.5 ± 0.1 mm
B = 0.3 mm minimum   E = 0.5 ± 0.1 mm
C = 0.8 – 1.1 mm

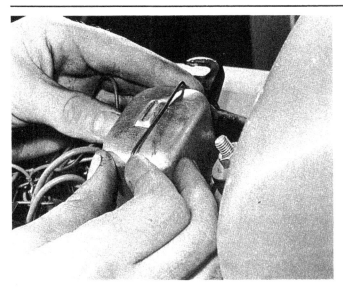

7.2a Remove regulator cover...

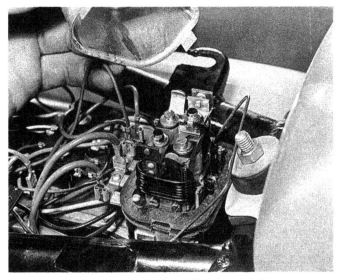

7.2b ...and examine contacts for signs of burning or pitting

little maintenance, except for the occasional cleaning of its contacts. If for any reason the regulator is suspected of being faulty it should be examined and tested as follows.

2   Disconnect the battery (negative terminal first), displace the wire clip which retains the regulator cover, and remove the cover and gasket. Visually inspect the regulator contacts for signs of pitting or burning. If the contacts are lightly pitted or burned they should be cleaned using a piece of fine emery cloth ensuring that their surfaces are left flat and are square to each other. However, if any set of contacts are badly pitted or burnt, the regulator can be considered faulty and should be renewed.

3   If all appears to be well, refer to the accompanying diagram and check all the contact gaps using a set of feeler gauges. If necessary, the gaps can be adjusted by carefully bending the springs. On no account bend the contacts themselves. Once the contacts are known to be clean and properly set, refit the gasket and cover and secure with its retaining clip. Reconnect the battery (negative terminal last).

4   Once all the regulator contacts are known to be clean and correctly gapped, check the generator output as described in Section 4 of this Chapter. If the alternator voltage is still not within the specified range, and all other charging system components are known to be in a good condition, the regulator is faulty and must be renewed.

8 Charging system warning lamp: function and location of faults

1   A warning lamp is fitted to give an early indication of any faults in the charging system. On models fitted with a tachometer this is the red lamp in the tachometer, and on models fitted with just a speedometer the warning lamp is the green lamp on the speedometer face. On these models the charging light serves a dual purpose, and is used as a turn signal warning lamp.

2   With the ignition switched on, but the engine stopped, the lamp will illuminate. The exception being on models with the combined charging/turn signal lamp where the lamp will flash on and off if the turn signals are switched on.

3   With the engine running and the charging system in good order the lamp should extinguish, except of course if the turn signals are switched on (models with a combined charging/turn signal lamp (in which case the lamp will flash on and off out of phase with the turn signal lamps when they are switched on)).

4   If the generator is not charging, the warning lamp will illuminate when the engine is run, to indicate a fault. On models with the combined charging/turn signal lamp, if the turn signals are switched on, the lamp will flash on and off with the turn signal lamps.

5   If the lamp fails to light when the ignition is switched on (engine

stopped), check the following items in the order given below:

(a)   Flat battery
(b)   Blown fuse
(c)   Blown bulb
(d)   A fault in either:
     1) The green/red wire from the warning lamp to terminal 61 on the regulator
     2) The green/blue wire (including the fuse) which links the DF terminals of the regulator and generator.
(e)   Faulty rectifier, test as described in Section 6.

6   If the lamp does not go out when the engine is running at higher speeds, check the generator output as described in Section 4 and proceed as instructed.

9 Fuses: location and renewal

1   The electrical system is protected by a total of four fuses which are situated in the fuse box behind the right-hand sidepanel. Always carry spare fuses of all the correct ratings and replace the spare fuses as soon as possible if they are ever used.

2   Before renewing a fuse that has blown, check that no obvious short

9.1 Fuse box is located behind right-hand sidepanel

circuit has occurred, otherwise the replacement fuse will blow immediately it is inserted.

3 If a fuse blows during a journey and no spare is available, a 'get you home' remedy is to remove the blown fuse, wrap it in silver paper and refit it in the fuse holder. The silver paper will restore the electrical continuity by bridging the broken fuse wire. This expedient should never be used if there is evidence of a short circuit or other major electrical faults, otherwise more serious damage will be caused. Replace the 'doctored' fuse at the earliest opportunity, to restore full circuit protection.

4 **Note**: *never use the above method to bridge the small fuse which is fitted to the green/blue wire which links the generator and regulator. This fuse must only be replaced by the specified 2AT (time delay) fuse*. In an emergency the machine can be used this fuse removed from the circuit. However, it should be noted that without this fuse the alternator will not charge the battery, hence the machine will be running off of battery power only, therefore limiting the amount of time which the machine can be used for.

10 Switches: examination and renovation

1 While the switches should give little trouble, they can be tested using a multimeter set to the resistance function or a battery and bulb test circuit. Using the information given in the wiring diagram at the end of this Manual, check that full continuity exists in all switch positions and between the relevant pairs of wires. When checking a particular circuit, follow a logical sequence to eliminate the source of the problem.

2 As a simple precaution always disconnect the battery (negative lead first) before removing any of the switches, to prevent the possibility of a short circuit. Most troubles are caused by dirty contacts, but in the event of the breakage of some internal part, it will be necessary to renew the complete switch.

3 It should, however, be noted that if a switch is tested and found to be faulty, there is nothing to be lost by attempting a repair. It may be that worn contacts can be built up with solder, or that a broken wire terminal can be repaired, again using a soldering iron.

4 While none of the switches require routine maintenance of any sort, some regular attention will prolong their life a great deal. In the author's experience, the regular and constant application of WD40 or a similar water-dispersant spray not only prevents problems occurring due to water-logged switches and the resulting corrosion, but also makes the switch much easier and more positive to use. Alternatively, the switch may be packed with a silicone-based grease to achieve the same result.

Ignition switch
5 To test the ignition switch, remove the cover from the bottom of the switch so that access can be gained to the switch terminals. Referring to the wiring diagram and using the colour-coded wires, check for continuity between the relevant terminals using a multimeter set to the resistance scale. If continuity exists between the relevant terminals in all switch positions, the switch can be deemed serviceable. However, if in one or more of the switch positions continuity does not exist between the specified terminals, the switch is faulty and must be renewed; no satisfactory repairs can be made.

Handlebar switches
6 The handlebar switch can be checked by tracing the wiring back to the headlamp and disconnecting it from the main wiring loom. Use the wiring diagram to identify the relevant wires of the switch to be tested, then check for continuity using the test equipment. The handlebar switch comprises three separate switches, all of which can be purchased separately if renewal is required.

7 Remove the handlebar switch assembly by slackening its two retaining screws and separating the two halves of switch. Old switches can be removed by unsoldering their connections, and the new components soldered into position. On refitting the switch to the handlebars pass the thinner of the two wiring cables over the bars, and the thicker cable under the bars and pass both cables out of the bottom of the switch. Reassemble the two halves of the switch, taking care not to trap the wiring, and tighten its retaining screws securely.

Neutral switch
8 Models which are equipped with a tachometer also have a neutral

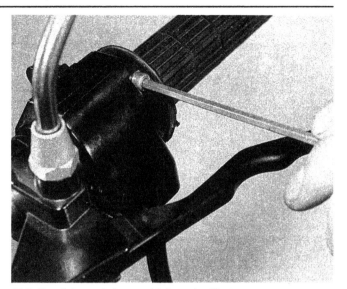

10.7a Handlebar switch assembly is retained by two screws

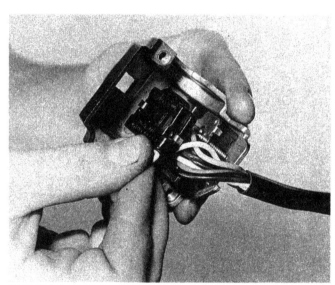

10.7b If faulty switches can be renewed individually

indicator lamp (yellow lamp situated in the tachometer). This lamp is operated by a switch fitted to the right-hand side of the crankcase where it is located just below the gearbox sprocket. Access to the switch may be gained once the right-hand crankcase cover has been removed (see Chapter 1).

9 To test the switch, set the multimeter to its resistance function and carry out a check for continuity between the switch terminal and earth with the transmission in neutral. If continuity is found, the switch is serviceable, but if not, the switch must be renewed. Note that although very unlikely, the fault may be due to the neutral switch contact which is fitted in the selector drum. This should be examined before condemning the switch. This contact is in the form of a rivet pressed into the selector drum, and is available as a separate item. Note that if the contact is damaged, it will be necessary to split the crankcase halves to gain access (see Chapter 1).

Front brake stop lamp switch
10 Models fitted with a front disc brake are also fitted with a front stop lamp switch, screwed into the master cylinder assembly. To test the switch, disconnect its wires, slacken its locknut and unscrew the switch from the master cylinder assembly. Check for continuity between the switch terminals. If the switch is serviceable there should be continuity when the spring-loaded plunger is out, and an open circuit (infinite resistance) when the plunger is pressed in. If this is not the case, the switch is faulty and must be renewed. On refitting, adjust the switch as described in Routine maintenance.

Rear brake stop lamp switch

11   All models are fitted with a rear stop lamp switch. The 251 models are equipped with a conventional plunger type switch linked to the brake pedal by means of a spring, whereas all other models have a switch which is an internal part of the rear brake assembly.

12   To test the switch on 251 models, remove the right-hand side-panel then trace the stop lamp switch wires back and disconnect them from the main wiring loom. If the switch is in good condition there should be continuity between the switch wires when the switch plunger is pulled down, and an open circuit (infinite resistance) when the plunger is released. If this is not the case, the switch is faulty and must be renewed.

13   On all other models, disconnect the wire from the switch terminal on the rear brake plate and check for continuity between the switch terminal and earth. If the switch is serviceable, there should be continuity when the brake is applied, and an open circuit (infinite resistance) when it is not. If not, try first to adjust the switch as described in Routine maintenance. If this fails the rear brake assembly must be removed and the stop lamp switch examined as described in Section 13 of Chapter 5.

11   Horn: location and testing

1   The horn is situated underneath the fuel tank where it is mounted on to the frame. No maintenance is required other than to remove road dirt and occasionally spray with WD40 or a similar water-dispersant fluid to minimise internal corrosion .

2   If the horn fails to work, first check that the battery is fully charged. Carry out a simple test to reveal whether current is reaching the horn by disconnecting the horn wires and substituting a 12 volt bulb. Switch on the ignition and press the horn button. If the bulb fails to light, check the horn button and wiring as described in Sections 2 and 10 of this Chapter. If the bulb does light, the horn circuit is proved good and the horn itself must be checked.

3   With the horn wires still disconnected, connect a fully charged 12 volt battery directly to the horn. If it does not sound, a sharp tap on the outside of the horn may serve to free the internal contacts. If this fails, the horn must be renewed as repairs are not possible.

4   Different types of horn may be fitted; if a screw and locknut is provided on the outside of the horn, the internal contacts may be adjusted to compensate for wear and to cure a weak or intermittent horn note. Slacken the locknut and slowly rotate the screw until the clearest and loudest note is obtained, then retighten the locknut. If no means of adjustment is provided on the horn fitted, it must be renewed.

12   Turn signal relay: location and testing

1   The turn signal relay is a sealed silver-coloured cylindrical unit located behind the right-hand sidepanel, next to the battery. On 125 and 150 models the relay is fitted inside a foam rubber band, and on 250, 251 and 300 models it is rubber-mounted via a plug on the top of the relay. These measures are necessary to protect the relay from vibration.

2   If the turn signal lamps cease to function correctly, there may be any one of several possible faults responsible which should be checked before the relay is suspected. First check that the lamps are correctly mounted and that all the earth connections are clean and secure. Check that the bulbs are of the correct wattage and that corrosion has not developed on the bulbs or their holders. Any such corrosion must be thoroughly cleaned off to ensure proper bulb contact. Also check that the turn signal switch is functioning correctly and that the wiring is in good order. Finally ensure that the battery is fully charged.

3   Faults in any one or more of the above items will produce symptoms for which the turn signal relay may be unfairly blamed. If the fault persists even after the preliminary checks have been made, the relay must be at fault. Unfortunately the only practical method of testing the relay is to substitute a known good one.

13   Bulbs: renewal

Headlamp and parking lamp

1   To renew the headlamp and/or parking lamp bulb remove the single

11.1 Location of horn

12.1 Turn signal relay is situated next to battery – 125 shown

retaining screw from the bottom of the headlamp and withdraw the headlamp unit. Unplug the three-pin headlamp connector and the parking lamp wire and lift the unit clear of the machine .

2   The bulbs are retained in the reflector by a spring clip; release the clip and withdraw the headlamp bulb or retainer plate complete with the parking lamp bulb. The parking lamp is a bayonet fit in the plate and can be removed by pressing it in and turning slightly to release it. The headlamp bulb can be simply lifted out of the reflector. On refitting make sure the tabs on the headlamp bulb are correctly positioned in the slot in the reflector. Check the headlamp beam settings, as described in Routine maintenance.

Instrument illuminating and warning lamps

3   All the instrument and warning lamp bulbs are fitted into holders which are pressed into the base of the instrument(s).

4   To gain access to these bulbholders it is necessary to displace the rubber instrument cover and slide it down the drive cable. The relevant bulbholder can then be unplugged and the bulb removed. All bulbs are a bayonet fit into the holders and can be removed by pressing in and turning anticlockwise. On refitting take care to ensure the rubber instrument cover is correctly seated .

Turn signals and stop/tail lamp

5   All turn signal lamp lenses and the stop/tail lamp lens are retained by screws. When removing take care not to tear the rubber seal which is

13.1a Remove screw from bottom of headlamp assembly...

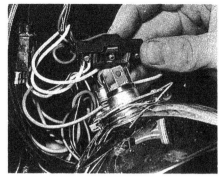

13.1b ...and unplug headlamp connector

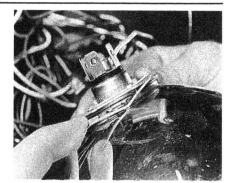

13.2a Release clip...

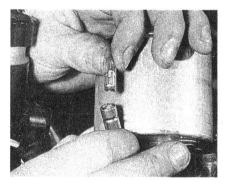

13.2b ...remove bulb retaining plate and parking lamp bulb...

13.2c ...and withdraw the headlamp bulb

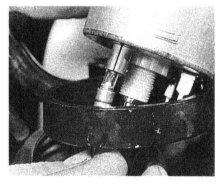

13.3 Bulbholders are a push fit into base of instrument...

13.4 ...and bulbs are a bayonnet fit in the holders

13.5a Turn signal lenses are retained by two screws...

13.5b ...and bulbs are a bayonnet fit in their holders – 251 shown

13.6a 251 models employ a combined stop/tail lamp bulb. Note offset pins on bulb...

13.6b ...whereas all other models use separate stop lamp...

13.6c ...and tail lamp bulbs

fitted between the lens and lamp. All bulbs are a bayonet fit in the lamps and can be removed by pressing in and turning anticlockwise.

6   On refitting, remove any moisture or corrosion from the bulb and holder and check that the contact presses firmly against the bulb. Note the offset pins on the combined stop/tail lamp bulb which is fitted to the 251 model. This prevents the bulb from being fitted incorrectly. Ensure the lens seal is correctly positioned then refit the lens taking care not to overtighten the retaining screws. Note that on all models except the 251 the front turn signal lamp lenses are fitted with projecting flanges which must be pointing upwards when the lens is fitted. These flanges act as an additional warning device to show the rider that the turn signals are functioning correctly.

Chapter 7 The 1991-on models

Contents

Specifications

Note: *The following specifications supersede those given in the preceding pages of this manual. Unless stated, the information applies to all models.*

Specifications relating to Routine maintenance

Engine

Spark plug:
Recommended grade	Isolator ZM 14-260, or NGK B8HS or B8HV
Electrode gap	0.6 mm (0.024 in)

Ignition timing:
Standard	2.5 + 0.5 mm (0.098 + 0.02 in) BTDC

Tolerance:
125 and 150 models	22°45' – 23°45' BTDC
251 and 301 models	20°15' – 22°15' BTDC

Cycle parts

Note: *Refer to tyre information label on motorcycle – it supersedes information given here*

Tyres pressures – tyres cold:	Front	Rear
125 and 150 models:		
Solo	1.5 kg/cm² (21 psi)	1.9 kg/cm² (27 psi)
Up to maximum permissible load	1.5 kg/cm² (21 psi)	2.8 kg/cm² (39 psi)
251 and 301 models:		
Solo	1.7 kg/cm² (24 psi)	1.9 kg/cm² (27 psi)
Up to maximum permissible load (3.25 x 16R tyre)	1.7 kg/cm² (24 psi)	2.8 kg/cm² (39 psi)
Up to maximum permissible load (110/80-16S tyre)	1.7 kg/cm² (24 psi)	2.5 kg/cm² (35 psi)

Specifications relating to Chapter 1

Engine

Capacity – 301 model	291 cc (17.8 cu in)
Compression ratio – 301 model	10:1

Cylinder barrel

Piston to bore clearance:
251 and 301 models	0.05 mm (0.002 in)

Specifications relating to Chapter 2

Fuel tank capacity

Saxon Fun models	23 lit (5.0 gal)
All other models	19 lit (4.2 gal)

BVF carburettor – early 301 model

Type	BVF 30 N 3-2
Main jet	135
Jet needle	2.5 B 511
Needle clip position – grooves from top	4th
Starter jet	95
Pilot jet	50
Pilot screw – turns out	2½
By-pass air screw – turns out	4

Bing carburettor

	125 and 150 models	**251 and 301 models**
Type	53/24/202 (125 cc) 53/24/201 (150 cc)	84/30/110
Main jet..	105	118
Needle jet..	66	72
Needle clip position – grooves from top	3rd (125 cc) 2nd (150 cc)	2nd
Pilot jet..	45	45
Pilot screw – turns out	1	1 – 1¼
Float height...	22.5 mm (0.89 in)	22.5 mm (0.89 in)
Idle speed..	1200 rpm	1200 rpm

Specifications relating to Chapter 3

Electronic ignition system

Ignition timing:
    Standard.. 2.5 + 0.5 mm (0.098 + 0.02 in) BTDC
    Tolerance:
        125 and 150 models....................................... 22º45' – 23º45' BTDC
        251 and 301 models....................................... 20º15' – 22º15' BTDC

Spark plug

Recommended grade .... Isolator ZM 14-260, or NGK B8HS or B8HV
Electrode gap .... 0.6 mm (0.024 in)

Specifications relating to Chapter 4

Rear suspension

Travel .... 135 mm (5.3 in)
Suspension unit spring free length... 270 + 8 mm (10.63 + 0.31 in)

Specifications relating to Chapter 5

Wheels

Front .... 1.85 x 18
Rear .... 2.15 x 16 (spoked), 2.50 x 16 (cast)

Front disc brake

Disc standard thickness .... Not available
Service limit* .... 3.5 mm (0.138 in)
*Markings stamped in disc mounting flange supersede information given here

Tyre sizes

Note: Refer to tyre information label on motorcycle – it supersedes information given here
Front:
    125 and 150 models .... 2.75 x 18R
    251 and 301 models .... 2.75 x 18R or 90/90-18S
Rear:
    125 and 150 models .... 3.25 x 16R
    251 and 301 models .... 3.25 x 16R or 110/80-16S

Specifications relating to Chapter 6

Fuses

Main .... 15A x 2
Turn signal 3A

Bulbs

Headlamp.. 12V 60/55W quartz halogen
Parking lamp.. 12V 5W

1 Introduction

This Chapter includes additional information to that in the preceding part of this Manual.

A number of significant changes were made in late 1991 with the move to electronic ignition, a Bing carburettor on certain models, and new electrical components. Noticeable styling changes included a new shape fuel tank (later to be of thermo-plastic construction), a new seat, tubular steel centre stand, plastic mudguards, plastic handlebar lever guards and restyled fairing on certain models. A new instrument panel replaced the separate units used on early models, and featured an electronically driven tachometer.

MZ produced an ETZ301 model in December 1991. Unlike the previous 300 cc model which was produced by the UK MZ importers using a cylinder barrel, head and piston sourced in the UK, the ETZ301 is a factory production model. It is identical to the ETZ251 model apart from having a different size cylinder liner and piston and benefits from the improvements described above.

Privatisation of the Motorradwerk Zschopau company followed the unification of Germany in 1992, and as a result the two-stroke model range underwent various stages of development. The Saxon models, introduced at the beginning of 1993 were the result of this development, the name originating from the German Saxony region. The 125 and 150 cc models took the name Saxon Roadstar or Saxon Sportstar, and 251 and 301 cc models were named Saxon Tour (code NE) or Saxon Fun (code VE).

Mechanical changes on the Saxon models were few, although all models underwent radical styling improvements and new electrical components and controls. Exact fitment is dependent on model type, but the later models can be identified by their new fairing, seat and bodypanels, cast alloy wheels, shorter silencer, folding footrests, square-bodied brake master cylinder, push-to-cancel turn signals, an engine kill switch and new QH square-shaped headlamp.

Where procedures or specifications for these later models differ from the main text, they are given in this supplementary chapter.

Note: *As a result of the company's privatisation and factory relocation, it has not been possible to tie all modifications down to a specific date or even model type, and owners are strongly advised to provide a full description of the items required when ordering parts.*

2 Routine maintenance

Revised schedule

Note: *Refer back to Routine maintenance at the beginning of this Manual for maintenance procedures, except where noted.*

Daily (pre-ride) checks

Check the engine oil level
Check the operation of the clutch and throttle grip
Check the operation of the brakes and the front brake fluid level (see below for square-bodied master cylinder)
Check the operation of the lights, turn signals, horn, speedometer and where fitted, the engine kill switch
Check the headlamp aim (see below for square-shaped unit)
Check the tyre pressures, tread depth and condition
Check the operation of the suspension and that there are no fluid leaks
Check the fuel level

Every 3100 miles (5000 km)

Check the transmission oil level
Check the fuel pipe condition
Clean the carburettor and check its settings
Check the oil pump cable adjustment
Clean the spark plug and check its gap
Check the battery electrolyte level, clean its terminals and check condition
Check the brake shoe and pad wear
Examine the wheels (see below for cast wheels)
Clean the fuel tap filters
Clean the air filter
Check and grease the drive chain
Lubricate the rear brake operating linkage
Lubricate the control lever pivots, stand pivot and throttle grip pulley

Fig. 7.1 Headlamp beam adjustment screws

1 Vertical adjustment screw
2 Horizontal adjustment screws

Every 6200 miles (10,000 km)

Check all fasteners for tightness, particulary the engine mountings
Renew the spark plug
Renew the air filter element
Lubricate the control cables and speedometer cable
Overhaul the drum brake, clean and lubricate the shoe pivots
Check the steering and suspension

Every 12,400 miles (20,000 km)

Change the transmission oil
Grease the wheel bearings and speedometer drive

Additional routine maintenance

Decarbonise the engine and exhaust system
Change the front brake fluid
Renew the front brake caliper and master cylinder seals
Renew the front brake hose
Clean the machine

2.2a Fluid level can be checked through sightglass in reservoir body

2.2b Remove reservoir cap, plate and diaphragm to add fluid

Front brake fluid level check – square-bodied master cylinder

1   A revised master cylinder with square-bodied fluid reservoir was introduced on later Saxon models.

2   The fluid level can be checked via the sightglass in the fluid reservoir body. If topping up is required, remove the two screws from the reservoir cap, lift off the cap, plate and rubber diaphragm. Top up the fluid as required. **Note:** *Take care to protect painted and plastic components from contact with brake fluid.*

3   The brake fluid type is marked on the reservoir cap.

Legal check – headlamp beam setting (square-shaped headlamp)

4   Headlamp beam setting adjustment on the square-shaped headlamp will require removal of the fairing (see Section 11).

5   Adjustment in the horizontal plane is made using the screw at each side of the unit, and vertical adjustment is made using the single screw at the base of the unit.

Cast alloy wheel examination

6   Check the complete wheel for cracks and chipping, particularly at the spoke roots and the edge of the rim. As a general rule, a damage wheel must be renewed as cracks will cause stress points which may lead to sudden failure under heavy load. Small nicks may be radiused carefully with a fine file and emery paper (No. 600 – No. 1000) to relieve the stress. If there is any doubt as to the condition of a wheel, seek the advice of an MZ dealer.

7   Each wheel is covered with a coating of lacquer or paint. If damage occurs to the wheel and the lacquer or paint finish is penetrated, the bared aluminium alloy will soon start to corrode. A whitish grey oxide will form over the damaged area. This should be removed and a new coating of protective lacquer or paint applied.

8   Note that no means is available of straightening a warped cast alloy wheel without resorting to the expense of having it skimmed on both faces; the safest action is to renew the wheel, especially if warpage has been caused as the result of an accident.

3   Engine: modification

Crankshaft left-hand oil seal renewal – 125 and 150 models

1   The crankshaft left-hand oil seal cap is not fitted to later engines. The new design oil seal locates directly in the crankcase and is fitted to all 125/150 models following frame serial numbers 4160371/4533500.

2   To renew the oil seal, separate the crankcases (see Chapter 1, Section 12) and remove the crankshaft from the left-hand casing (see Chapter 1, Section 13). Withdraw the two main bearings from the left-hand casing, then remove the circlip. The oil seal can be drifted or levered out of the crankcase.

3   Tap the new oil seal squarely into the casing from the outside and install the circlip. **Note:** *Use a tubular drift which bears only on the seal's hard outer edge.* Install the main bearings, refit the crankshaft and join the crankcase halves (see Chapter 1, Sections 28 through 30).

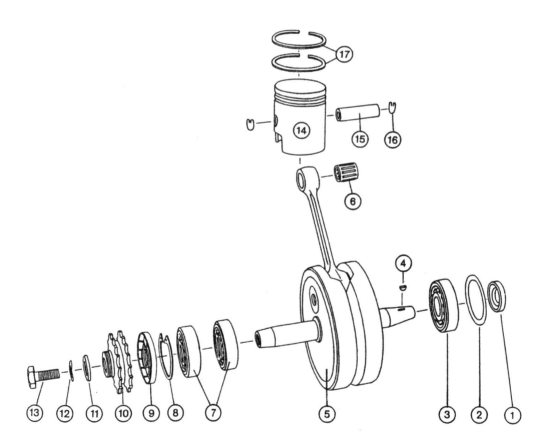

Fig. 7.2 Crankshaft and piston – 125 and 150 models (Sec 3)

1 Oil seal	5 Crankshaft	8 Circlip	13 Bolt
2 Shim	6 Small-end bearing	9 Oil seal	14 Piston
3 Right-hand main bearing	7 Left-hand main bearing –	10 Primary drive sprocket	15 Gudgeon pin
4 Woodruff key	2 off	11 Washer	16 Circlip – 2 off
		12 Spring washer	17 Piston rings

4 Fuel tank: modifications – Saxon models

Examination – plastic tank

1   The fitting of a plastic fuel tank was introduced on the Saxon models. Whilst removal and refitting of the tank remain as described in Chapter 2, Section 2, tank repair will require a different technique.
2   If the tank is leaking or has suffered accidental damage, seek the advice of an MZ dealer or specialist in this type of work.

Removal and refitting – Saxon 251 and 301 Fun models

3   Remove the two screws, with nylon washers from the joint between the fairing and fuel tank. Remove the tank as described in Chapter 2, Section 2.

5 Fuel tap: modifications – Saxon models

1   The fuel tap on later models is retained to the fuel tank by two bolts and spring washers.
2   When removing and refitting the fuel tap, check the condition of the O-ring between the tap body and tank; if flattened or deteriorated it must be renewed.

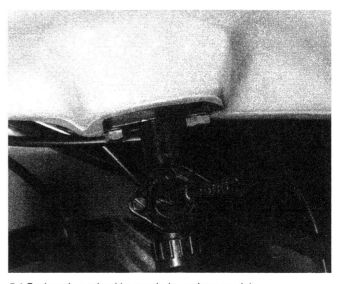

5.1 Fuel tap is retained by two bolts on later models

Fig. 7.3 Bing carburettor – 125 and 150 models (Sec 6)

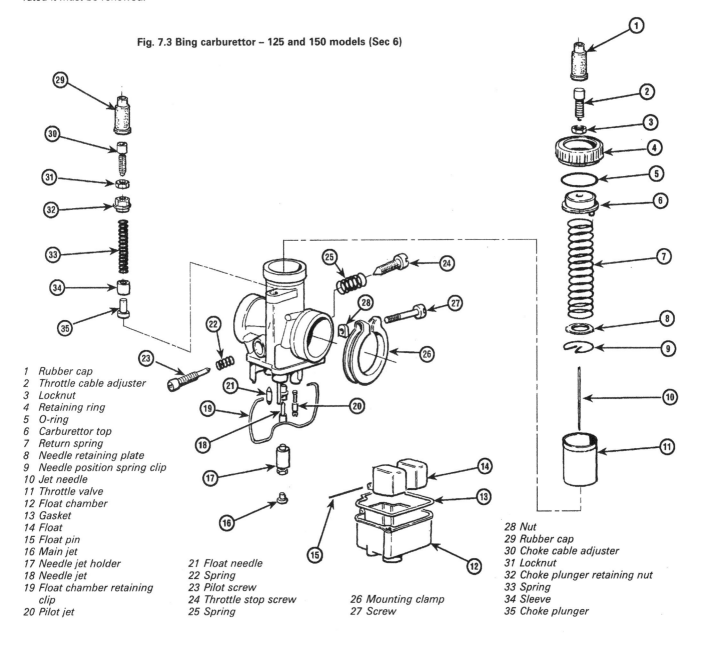

1   Rubber cap
2   Throttle cable adjuster
3   Locknut
4   Retaining ring
5   O-ring
6   Carburettor top
7   Return spring
8   Needle retaining plate
9   Needle position spring clip
10   Jet needle
11   Throttle valve
12   Float chamber
13   Gasket
14   Float
15   Float pin
16   Main jet
17   Needle jet holder
18   Needle jet
19   Float chamber retaining
       clip
20   Pilot jet

21   Float needle
22   Spring
23   Pilot screw
24   Throttle stop screw
25   Spring

26   Mounting clamp
27   Screw

28   Nut
29   Rubber cap
30   Choke cable adjuster
31   Locknut
32   Choke plunger retaining nut
33   Spring
34   Sleeve
35   Choke plunger

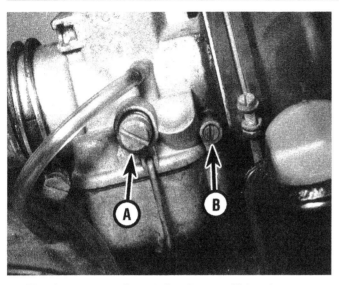

6.3 Throttle stop screw (A) and pilot air screw (B) locations – 251/301 models

7.2 Silencer rear mounting bolts to footrest bracket on later models

6 Bing carburettor: general information

1   Later models are fitted with a Bing carburettor instead of the BVF component fitted previously.
2   Refer to the Specifications section of this Chapter for jet sizes and settings.
3   Follow the procedures in Chapter 2 for removal/refitting, overhaul and adjustment of the carburettor. The Bing carburettor fitted to 125 and 150 models is shown in the accompanying illustration. On 251 and 301 models the carburettor components conform to the assembly shown in Fig. 2.3 of Chapter 2, although note that the throttle stop screw and pilot screw are located on the left-hand side of the carburettor.

7 Exhaust system: modified rear mounting

1   On Saxon models fitted with the new style footrest, the exhaust system has a revised rear mounting which dispenses with the long mounting rod used on all other models.
2   Remove and refit the exhaust system as described in Chapter 2, Section 10, noting that the silencer is bolted to an extension of the right-hand footrest bracket. A rubber bush is housed in the footrest bracket and the mounting is formed by a bolt, plain washer, collar (inserted from the inner side) and nut.

8 Ignition system: electronic components

Description
1   Later models use an electronic ignition system, dispensing with the need for mechanical contact breaker points. A sensor mounted on the generator stator is triggered by a rotor on the end of the crankshaft. As the rotor sweeps past the sensor, a pulse is transmitted to the ignition control unit which discharges the current through the ignition coil primary winding.

Fault diagnosis
2   Due to the system having no mechanical parts it should remain trouble-free. Failure is likely to be due to a wiring fault, or total failure of a single component.
 a) *Loose, corroded or damaged wiring connections, broken or shorted wiring between any of the ignition system components*
 b) *Faulty ignition switch. Examine switch as described in Chapter 6*
 c) *Faulty ignition HT coil*
 d) *Faulty ignition control unit or sensor.*

Sensor and rotor location
3   Remove the four screws and detach the crankcase right-hand cover. **Note:** *On models equipped with the cast footrest brackets, removal of the cover will be made easier if the bracket is withdrawn*

8.4a Ignition sensor/control unit is retained by two screws

8.4b Release its wiring from tab to free sensor/control unit

8.5 Generator bolt retains ignition rotor

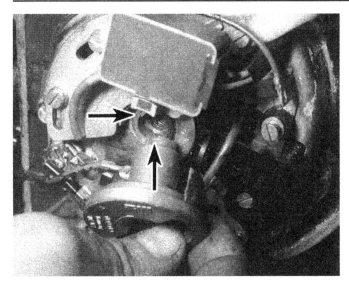

8.6 Slot in ignition rotor must align with peg on refitting (arrowed)

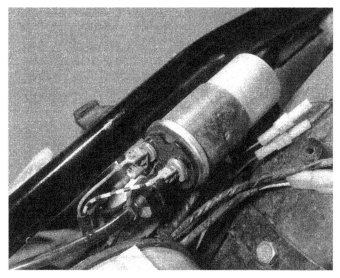

8.9 Ignition HT coil is located under the seat

slightly (see Section 9); take care not to damage the delicate regulator/rectifier unit when removing the cover. The sensor is mounted on the generator stator plate by two screws. Before removing the screws make alignment marks across the stator plate and generator stator edges as a reference on refitting; these slots provide provision for adjustment of the ignition timing.

4   To free the sensor, remove the two retaining screws and free its wiring from the clamp. Trace the wires up to the ignition coil and disconnect them.

5   The ignition rotor is retained to the crankshaft by the long generator rotor retaining bolt. Remove the bolt and spring washer and withdraw the rotor. **Note:** *The bolt should release with a sharp tap on the end of a long ring spanner, but if not lock the engine through the transmission to hold the crankshaft stationary.*

6   The ignition rotor is keyed to the crankshaft end by slot and peg arrangement; ensure these align on refitting.

7   The first electronic ignition equipped models to appear in the UK had separate sensor and control units, as opposed to the combined unit described above. Note that it is important to maintain an air gap of 1.2 to 1.6 mm (0.047 to 0.063 in) between the sensor transmitter and the rotor magnet on the early system.

Ignition control unit

8   The ignition control unit is integral with the sensor (see above). Note, however, that some early UK models had separate units; with the control unit being mounted behind the right-hand sidepanel or under the seat.

Ignition HT coil

9   The coil is mounted under the seat, on the frame rail.

9   Footrests and rear brake pedal: removal and refitting – later models

Footrests

1   The footrests are of the folding type, and are supported on cast brackets.

2   To remove a footrest, prise the E-clip and washer free, then withdraw the clevis pin. Withdraw the footrest and washer from the bracket.

Footrest brackets

3   Remove the silencer (see Section 7). Remove the throughbolt from the base of the footrest bracket; it passes through the brake pedal crank and is retained by a nut and washer on its inner side. Remove the swinging arm right-hand side locknut, nut and washer and withdraw the footrest bracket off the swinging arm pivot shaft end. Note that the brake link rod must be detached from the brake pedal.

9.3 Unhook the rear brake link rod from the pedal eye

9.5 Brake pedal is retained to footrest bracket by throughbolt and nut

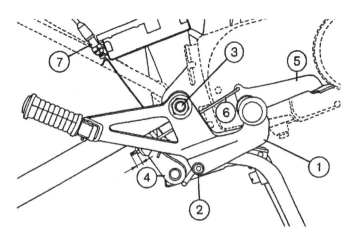

**Fig. 7.4 Right-hand footrest bracket and brake pedal detail —
later models (Sec 9)**

1 Footrest bracket	*4 Brake pedal crank*
2 Throughbolt	*5 Brake pedal*
3 Swinging arm pivot shaft	*6 Brake pedal link rod*
nuts	*7 Stop lamp switch*

4   The left-hand bracket is retained by the swinging arm pivot nut and by the brake pedal cross-over shaft lever throughbolt, washer and nut.

Rear brake pedal and linkage
5   The brake pedal pivots about the rider's footrest mounting. To remove the pedal, remove the right-hand footrest bracket (see Step 3), and detach the rider's footrest (see Step 2). The pedal and footrest mount can be separated after removal of the throughbolt and nut. On refitting, apply a smear of grease to the bearing surfaces of the brake pedal and nut collar before refitting the throughbolt and nut.
6   The brake pedal height, in relation to the rider's footrest, can be adjusted via the locknut and adjuster screw at the pivot end of the pedal, and by the adjuster nut at the end of the pedal link rod.
7   The manufacturer advises a standard setting of 10 mm exposed thread at the link rod end (see the accompanying illustration). After adjustment, check the operation of the rear brake stop lamp switch, and if necessary remove the right-hand side panel and adjust the switch height accordingly.

10 Speedometer and tachometer heads: removal and refitting

Individual meters
1   The meters can be removed independently of each other. First remove the fairing (see Section 11).

2   Start by easing the meter illuminating bulbholder out of the base of the meter.
3   In the case of the speedometer, unscrew the knurled ring and withdraw the cable from the base of the meter.
4   Remove the two cap nuts with collars and lift the meter off the mounting base until the wiring can be detached from its base. To free the tachometer, carefully disconnect its wire connector plug. To free the speedometer, pry the bulbholders for the warning lights out of their locations.
5   The meters are sealed units and cannot be repaired. **Note:** *Always store the meters the correct way up when removed; failure to do so may result in damage to their delicate movement.*
6   Refitting is a reverse of the removal procedure.

Complete instrument panel
7   Remove the fairing (see Section 11).
8   Trace all wiring from the instruments and disconnect it at the connectors. Unscrew its knurled ring and withdraw the speedometer cable.
9   The instrument panel is mounted on brackets which extend from the fork legs. Pull off the plastic covers from the two mounting nuts, remove the nuts, washers and damping rubbers and lift the instruments off the fork leg brackets (see photo 10.4a).
10   When refitting the instruments, take care to ensure the damping rubbers are fitted correctly, they protect the meters from vibration.

11 Fairings and bodywork: removal and refitting

Fairing – all basic models and Saxon 125 Roadstar
1   The fairing is retained on its mounting brackets by four screws.

Fairing – Saxon 125 Sportstar and 251/301 Tour models
2   Remove the two screws from the underside of the fairing. Reach inside the fairing and disconnect the wiring to the turn signals, then remove the single screw which retains each turn signal lamp to the fairing and withdraw the wiring.
3   Refit in a reverse of removal, noting that the turn signal wiring must pass through the square aperture in the mounting bracket.

Fairing – Saxon 251/301 Fun models
4   Remove the two screws, with nylon washers from the joint between the fairing and fuel tank. Reach inside the fairing and disconnect the wiring to the turn signals, then remove the single screw and nut which retain each turn signal lamp to the fairing and withdraw the wiring.
5   Refit in a reverse of removal, noting that the turn signal wiring must pass through the square aperture in the mounting bracket.

Tail fairing
6   Remove the two nuts from the base of the seat, then disengage it from its front hooks, lifting the seat clear.
7   Remove the four screws from the plastic bridging panel beneath the grab rail and lift it free.
8   Trace and disconnect the wiring from the rear turn signals at the

10.3 Unscrew its knurled ring and detach speedometer cable

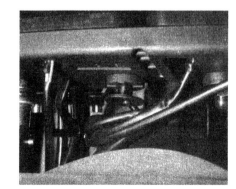

10.4a Individual meters are retained by two cap nuts (A). Note instrument base mounting nut (B)

10.4b Instrument mounting collars pass through rubber mounting grommets

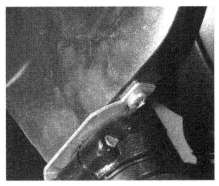

11.1a On basic models, fairing is retained by a single screw on each lower edge ...

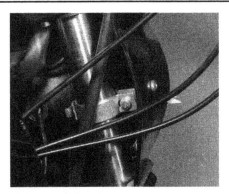

11.1b ... and by a single screw at each side (note the mounting clamp arrangement on the fork leg)

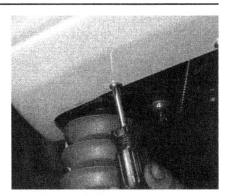

11.2 Fairing lower retaining screws on Tour models

11.3 Turn signal wiring passes through aperture in mounting bracket (arrowed)

11.7 Tail fairing bridging panel is retained by four screws

11.8 Remove nut (arrowed) to release turn signal from tail fairing

11.9 Tail fairing front mounting

11.10a Hand guards are retained by the mirror mounting ...

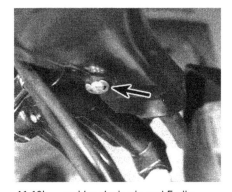

11.10b ... and by clevis pin and E-clip (arrowed)

connectors. Unscrew the nut from the back of the turn signal mountings and withdraw the turn signal and its wiring.

9   Remove the Torx screw from the suspension unit upper mounting and lift the tail fairing left and right sections away.

Hand guards

10   The plastic hand guards are retained by the mirror mounting and by a clevis pin and E-clip.

12 Hydraulic braking system: modifications

General modifications

1   A slimmer, lighter caliper is fitted to later models, which uses sintered metal pads. The brake disc thickness is reduced and its braking surface is slotted for improved heat dissipation.

2   Master cylinder piston diameter is reduced from 12.7 mm to 11.5 mm.

Brake disc mounting

3   The front brake disc is retained to the wheel hub by six Torx-head screws on later models. Note that these screws can be removed and refitted a maximum of four times, whereupon they must be renewed.

4   Note that the disc thickness service limit is stamped in the mounting flange – this information supersedes that given in the Specifications.

Brake master cylinder

5   A new Grimeca master cylinder is fitted to certain later models, identified by its square-bodied fluid reservoir. Its mountings are the same as the earlier type unit described in Chapter 5, Section 9. No

12.4 Service limit is stamped in brake disc flange

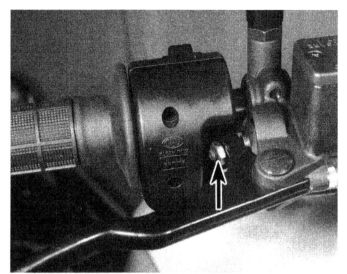

12.6 Brake lever position is adjustable via screw and locknut (arrowed) on later master cylinder

details are available of the piston assembly – if this requires attention, seek the advice of an MZ dealer.

6   The new master cylinder provides a means of brake lever position adjustment. To adjust, slacken the locknut and alter the setting of the adjuster screw in the lever; tighten the locknut on completion. Check that the front brake stop lamp switch functions correctly after adjustment.

13 Rear wheel sprocket and cush drive: modification

1   A separate rear wheel sprocket and cush drive hub replace the combined component fitted to earlier models. Remove the sprocket and cover assembly as described in Chapter 5, Section 7.

2   To separate the sprocket from the hub, prise out the cush drive rubber damper to reveal the sprocket bolt retaining nuts. Flatten back the tabs on the retaining bolts and unscrew them.

3   When fitting the sprocket, secure the bolt heads by bending up an unused portion of the lock washers; if necessary, renew the lock washers.

14 Regulator/rectifier unit: general information

1   On later models the separate voltage regulator and rectifier units are replaced by a combined unit, mounted on the generator stator.

2   No test details are available for the combined unit. If a fault develops in the charging system refer to an MZ dealer.

15 Fuses: location and renewal

1   Later models are fitted with two 15A plug-in type fuses, mounted in a holder next to the battery. Remove the right-hand side panel for access.

2   Unclip the transparent fuse cover and pull the fuses out of position using your fingers or tweezers. The fuse rating is clearly marked on the plastic fuse backing. The fuse holder has provision for two spare fuses.

3   A blown fuse will have a broken link between its two spade terminals. Always trace and rectify the cause of the problem first before renewing a blown fuse.

4   Certain models may have an additional fuse for the turn signal circuit, located either in the fuse box or as a separate cartridge-type fuse in an in-line holder.

16 Switches: modification

Handlebar switches

1   New design handlebar switches were fitted to all later models. The right-hand switch is retained by two screws at the front, and the left-hand switch is retained by three screws on the underside.

2   The switches can be tested for continuity as described in Chapter 6 in conjunction with the relevant wiring diagram at the end of this chapter.

Ignition switch

3   The switch is mounted to the underside of the top yoke by two bolts. Note that the ignition switch doubles as the lighting switch on the lower specification models.

4   If renewal is required, remove the fairing (see Section 11), and instruments (see Section 10). Disconnect the wiring for the switch and remove its two bolts to free it from the top yoke.

5   Continuity checks can be made either at the switch wire connector block, or directly across its terminals once the lower cover has been unscrewed from the switch body. Refer to the relevant wiring diagram at the end of this chapter and, using a multimeter set to the resistance scale, make continuity checks in all switch positions.

14.1 Combined regulator/rectifier unit – later models

17.1 Turn signal relay is mounted on tool tray – later models

18.3 Headlamp mounting frame attaches to lower yoke by two throughbolts and nuts

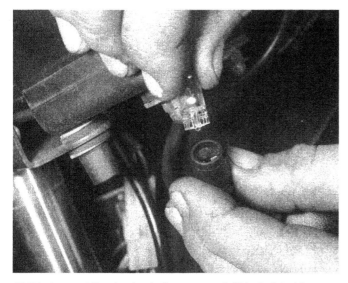

19.2 Instrument illuminating bulbs are a push fit in their holders

17 Turn signal relay: modification

Later models are equipped with a modified relay, located in a rubber mounting tab fixed to the tool tray. Remove the right-hand side panel for access.

18 Headlamp: bulb renewal

Round headlamp unit

1   Refer to the procedure given in Chapter 6, Section 13, noting that the headlamp bulb is of the quartz halogen type. **Caution:** *When installing the new bulb take care not to touch the bulb's glass portion with your fingers – oil from your skin will cause the bulb to overheat and fail prematurely.*

Square headlamp unit

2   Remove the fairing (see Section 11).
3   Due to restricted access to the back of the headlamp unit it is

recommended that the headlamp mounting frame be detached from its mountings and withdrawn fully forwards. The frame is retained by two throughbolts to the lower yoke and by a bolt each side to the brackets at the fork tops.
4   As the mounting bracket is withdrawn, peel back the rubber dust cover, then unplug the wiring connector from the headlamp and pull out the parking lamp bulb.
5   Peel back the large rubber dust cover and rotate the plastic retainer ring anti-clockwise to free the headlamp bulb. **Caution:** *When installing the new bulb take care not to touch the bulb's glass portion with your fingers – oil from your skin will cause the bulb to overheat and fail prematurely.* Secure the bulb by rotating the plastic retainer into its slots in the unit.
6   The parking lamp bulb is of the capless type and is simply pulled out of its bulbholder.
7   Plug the wire connector onto the headlamp bulb terminals and push the parking lamp bulbholder into the reflector. Refit the dust cover over the back of the headlamp, engaging its cutout with the tab.
8   Tighten the headlamp mounting frame bolts securely.

19 Instruments: bulb renewal and electronic tachometer

Bulb renewal

1   Remove the fairing for access to the instruments (see Section 11).
2   The speedometer and tachometer illuminating bulbs can be accessed by carefully prying out the rubber bulbholders from the base of the meters. The bulbs are of the capless type and can be removed by carefully pulling them out of the bulbholder.
3   The warning lamp bulbs housed in the speedometer, can only be accessed after the meter has been lifted off the mounting bracket. Pull the illuminating bulbholder from the base of the meter and unscrew the speedometer cable. Remove the two cap nuts and lift the meter up as much as the wiring will allow without straining it.
4   Pull the warning lamp bulbholders from position in the base of the meter. The capless bulbs can be pulled out of their holders.

Electronic tachometer

5   The tachometer operates off the ignition system. If it fails, first check that the wiring and connectors are secure. No test details are available for the meter; if failure cannot be traced to the wiring or connections the tachometer must be renewed.

Key to wiring diagram components on pages 127 to 130

1	Battery
2	Fuses
3	Ignition and lighting switch
4	Regulator
5	Rectifier
6	Generator
7	Neutral switch
8	Neutral warning lamp (where fitted)
9	Charging lamp (also turn signal warning lamp on standard models)
10	Horn switch
11	Headlamp pass switch
12	Headlamp dip switch
13	Main beam warning lamp
14	Headlamp main beam
15	Headlamp dip beam
16	Tachometer lamps (where fitted)
17	Speedometer lamps
18	Parking lamp
19	Tail lamp bulb*
20	Stop lamp bulb*
21	Spark plug
22	Ignition HT coil
23	Contact breaker and condenser
24	Turn signal relay
25	Front brake stop lamp switch
26	Rear brake stop lamp switch
27	Turn signal switch
28	Turn signal warning lamp (Luxus models)
29	Front left-hand turn signal
30	Rear left-hand turn signal
31	Front right-hand turn signal
32	Rear right-hand turn signal
33	Connections for models with sidecar (where applicable)
34	Capacitor
35	Horn

*stop/tail lamp combined on ETZ251 model

Key to symbols

MA	Earth point in headlamp housing
MB	Earth point of stop/tail lamp
ML	Earth for headlamp
MC	Main earth connection
MD	Generator earth
MT	Speedometer earth
bl	blue
br	brown
ge	yellow
gn	green
gr	grey
rt	red
sw	black
ws	white

Notes

1	Wire size is given in millimetres after colour code.
2	Wires marked 'A' and shown in dash and dot form relate to the standard models which are not fitted with a tachometer.
	Wires marked 'B' and shown in dash form relate to Luxus models, fitted with a tachometer.

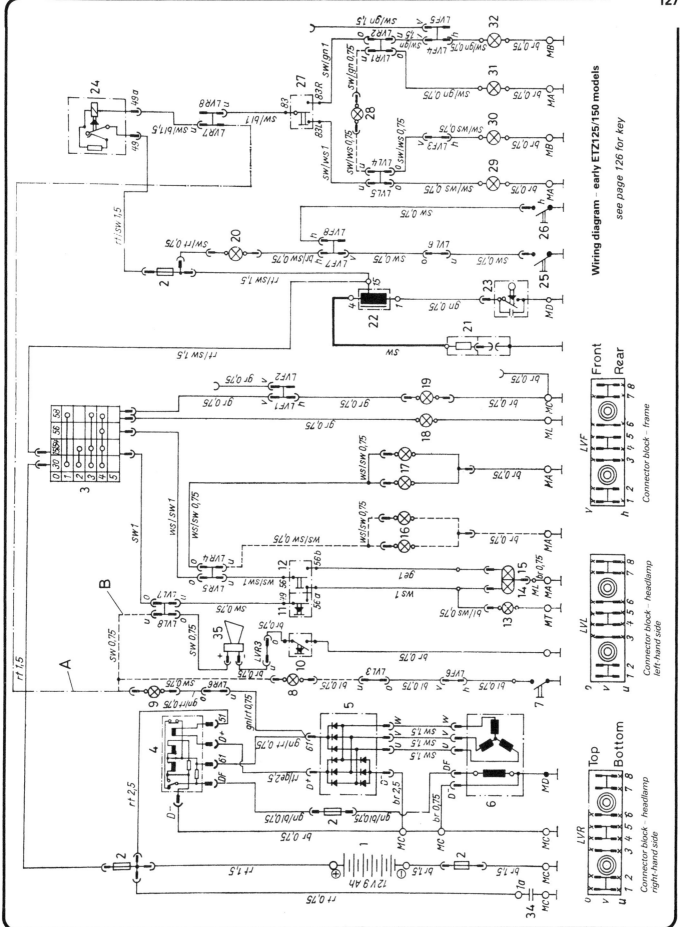

Wiring diagram – early ETZ125/150 models

see page 126 for key

Connector block – frame

LVF

Front / Rear

Connector block – headlamp left-hand side

LVL

Connector block – headlamp right-hand side

LVR

Top / Bottom

128

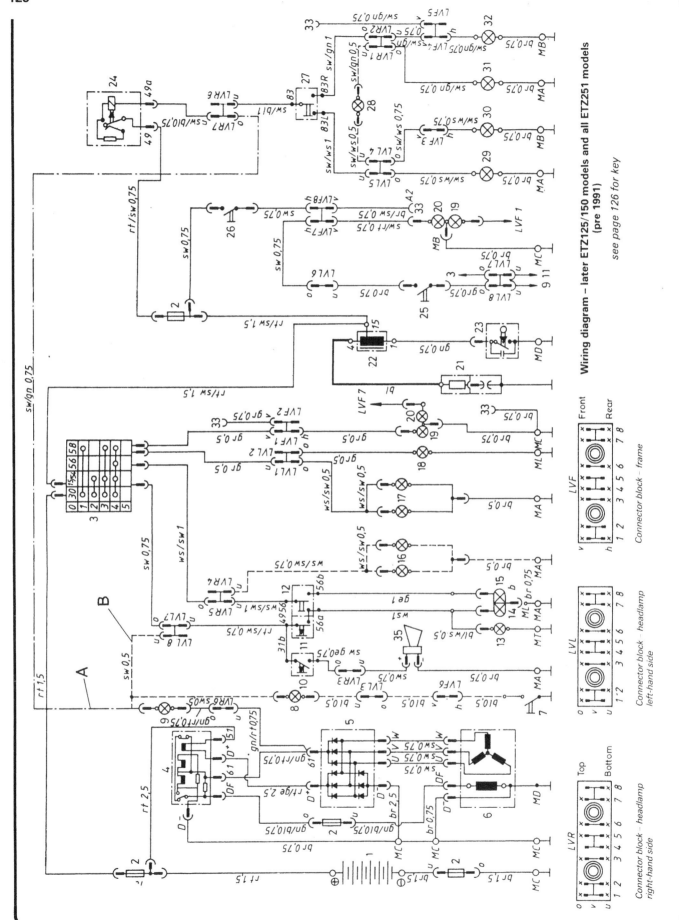

Wiring diagram – later ETZ125/150 models and all ETZ251 models (pre 1991)

see page 126 for key

Connector block – frame

Connector block – headlamp left-hand side

Connector block – headlamp right-hand side

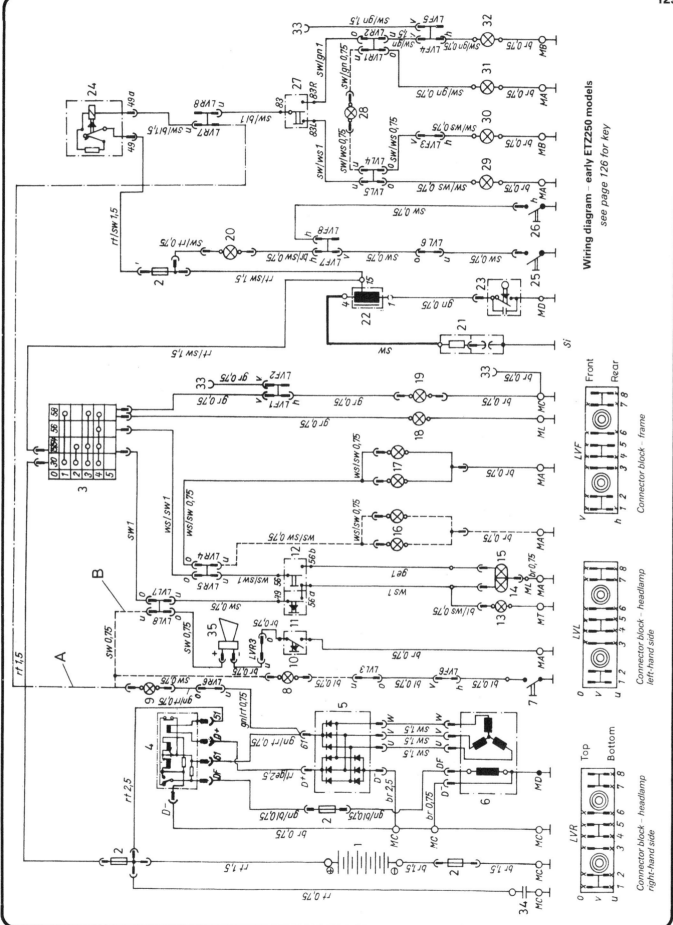

Wiring diagram – early ETZ250 models

see page 126 for key

Connector block – frame

Connector block – headlamp
left-hand side

Connector block – headlamp
right-hand side

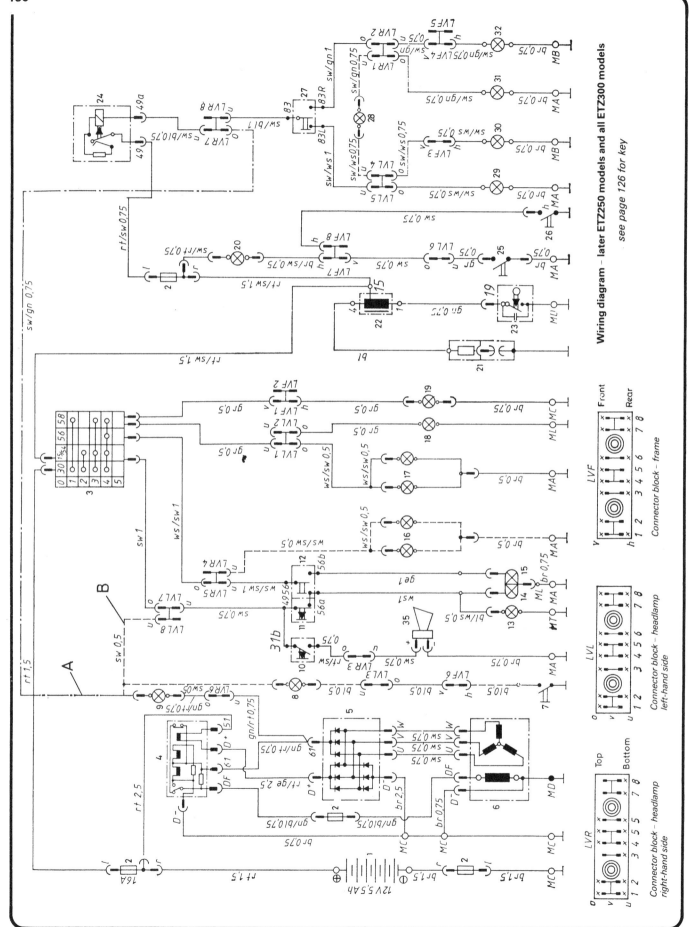

Wiring diagram – later ETZ250 models and all ETZ300 models

see page 126 for key

Connector block – frame

Connector block – headlamp left-hand side

Connector block – headlamp right-hand side

Key to wiring diagram components on page 132

1   Battery
2   Generator
3   Regulator/rectifier unit
4   Electronic tachometer
5   Charging warning lamp
6   Neutral lamp
7   Neutral switch
8   Horn switch
9   Horn
10 Headlamp flash switch
11 Headlamp dip switch
12 High beam warning lamp

13 Headlamp
14 Instrument illuminating lamp
15 Instrument illuminating lamp
16 Parking lamp
17 Tail lamp
18 Ignition HT coil
19 Contact breaker and condenser*
20 Spark plug
21 Front brake stop lamp switch
22 Rear brake stop lamp switch
23 Brake stop lamp
24 Turn signal relay (early type)

25 Turn signal switch
26 Turn signal warning lamp
27 Rear left-hand turn signal
28 Front left-hand turn signal
29 Front right-hand turn signal
30 Rear right-hand turn signal
32 Electronic ignition system
33 Control unit
34 Sensor
35 Turn signal relay (later type)

*Not fitted to UK models

Key to wiring diagram components on pages 133 and 134

1   Headlamp
1a Parking lamp
1b Headlamp dip beam
1c Headlamp main beam
2   Speedometer
2a Turn signal warning lamp
2b High beam warning lamp
2c Charging warning lamp
2d Neutral lamp
2e Illuminating lamp
3   Tachometer
3a Illuminating lamp
3b Electronic tachometer
4   Left-hand handlebar switch

4a Horn switch
4b Headlamp dip switch
4c Headlamp flash switch
4d Turn signal switch
4e Lighting switch
5   Ignition switch
6   Front left-hand turn signal
7   Front right-hand turn signal
8   Horn
9   Ignition HT coil
10 Spark plug
11 Generator assembly
11a Generator
11b Electronic ignition sensor/control
    unit

11c Regulator/rectifier unit
12 Turn signal relay
13 Fuses
14 Battery
15 Stop/tail lamp
15a Stop lamp
15b Tail lamp
16 Rear left-hand turn signal
17 Rear right-hand turn signal
18 Rear brake stop lamp switch
19 Front brake stop lamp switch
20 Neutral switch
21 Turn signal circuit fuse
22 Engine kill switch
23 Diode

Key to symbols

bl	blue
br	brown
ge	yellow
gn	green
gr	grey
rt	red
sw	black
ws	white

Note: *Wire size is given in millimetres after colour code*

132

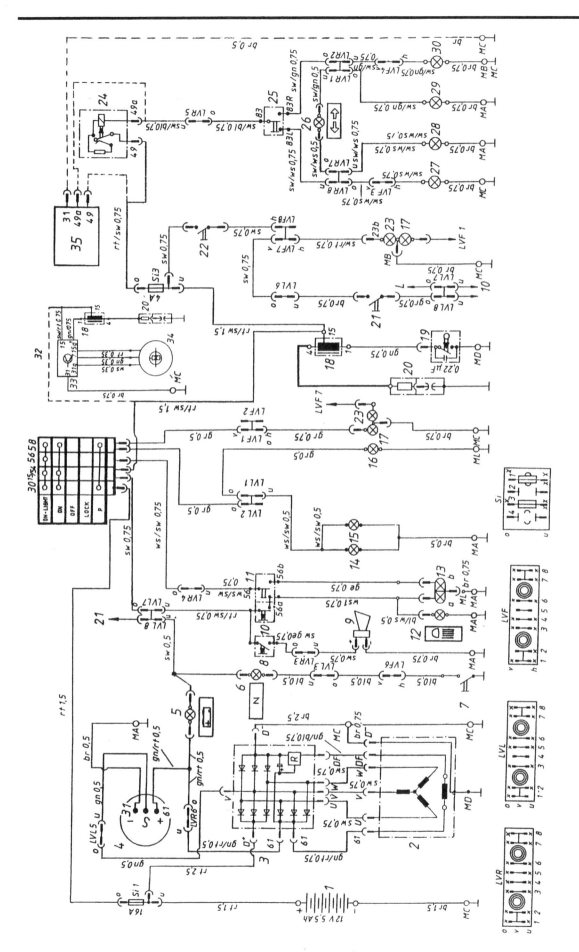

Wiring diagram — late 1991-on ETZ125, 150, 251 and 301 Luxus models

see page 131 for key

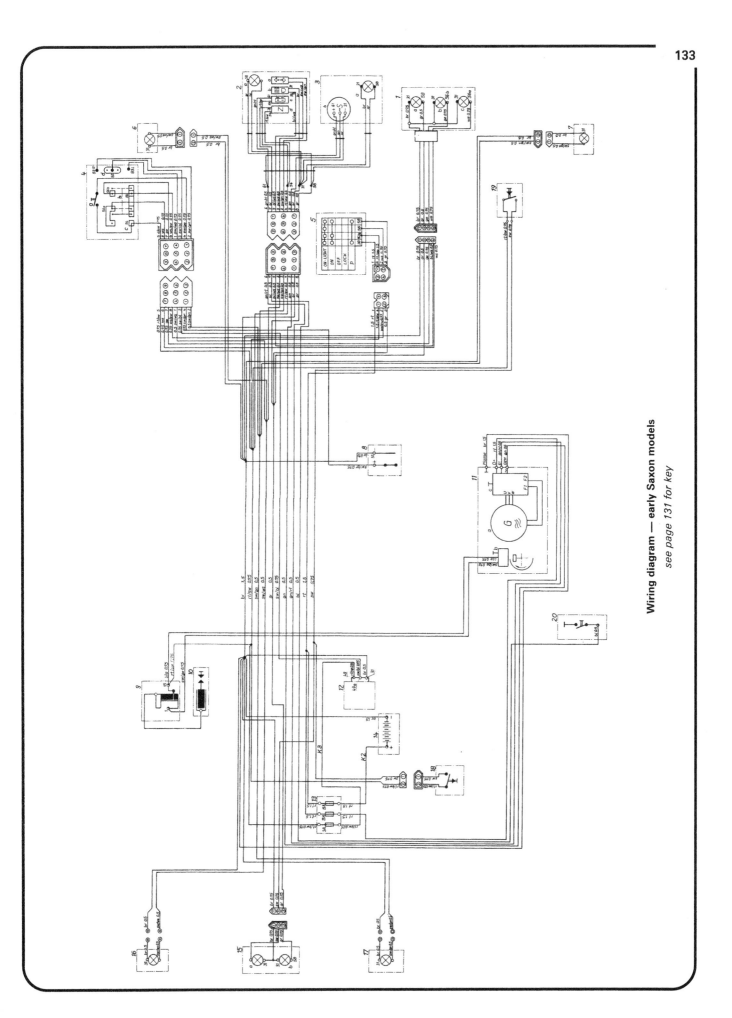

Wiring diagram — early Saxon models
see page 131 for key

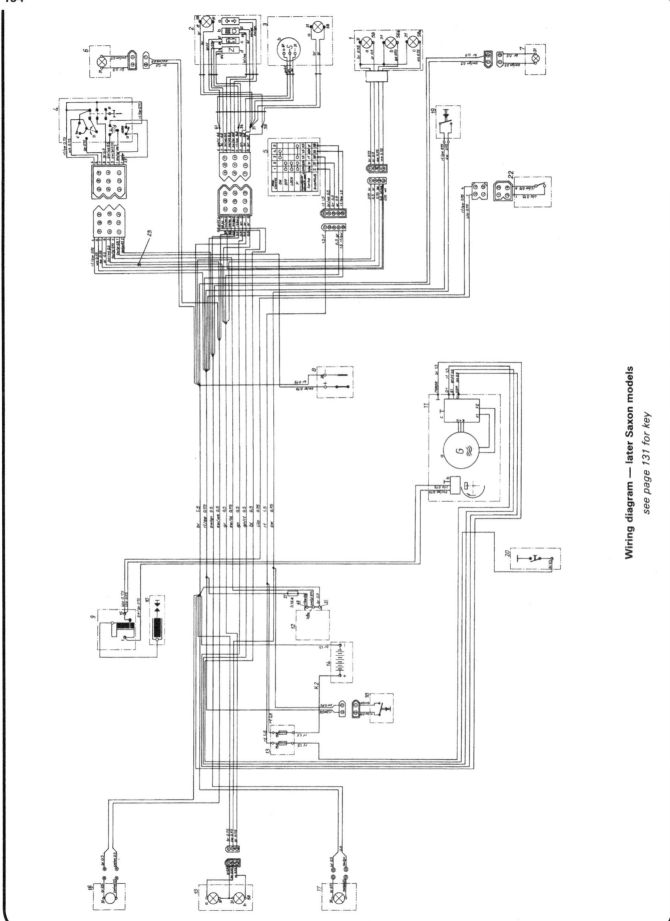

Wiring diagram — later Saxon models
see page 131 for key

Conversion factors

Length (distance)

Inches (in)	X	25.4	= Millimetres (mm)	X	0.0394	= Inches (in)	
Feet (ft)	X	0.305	= Metres (m)	X	3.281	= Feet (ft)	
Miles	X	1.609	= Kilometres (km)	X	0.621	= Miles	

Volume (capacity)

Cubic inches (cu in; in³)	X	16.387	= Cubic centimetres (cc; cm³)	X	0.061	= Cubic inches (cu in; in³)	
Imperial pints (Imp pt)	X	0.568	= Litres (l)	X	1.76	= Imperial pints (Imp pt)	
Imperial quarts (Imp qt)	X	1.137	= Litres (l)	X	0.88	= Imperial quarts (Imp qt)	
Imperial quarts (Imp qt)	X	1.201	= US quarts (US qt)	X	0.833	= Imperial quarts (Imp qt)	
US quarts (US qt)	X	0.946	= Litres (l)	X	1.057	= US quarts (US qt)	
Imperial gallons (Imp gal)	X	4.546	= Litres (l)	X	0.22	= Imperial gallons (Imp gal)	
Imperial gallons (Imp gal)	X	1.201	= US gallons (US gal)	X	0.833	= Imperial gallons (Imp gal)	
US gallons (US gal)	X	3.785	= Litres (l)	X	0.264	= US gallons (US gal)	

Mass (weight)

Ounces (oz)	X	28.35	= Grams (g)	X	0.035	= Ounces (oz)
Pounds (lb)	X	0.454	= Kilograms (kg)	X	2.205	= Pounds (lb)

Force

Ounces-force (ozf; oz)	X	0.278	= Newtons (N)	X	3.6	= Ounces-force (ozf; oz)
Pounds-force (lbf; lb)	X	4.448	= Newtons (N)	X	0.225	= Pounds-force (lbf; lb)
Newtons (N)	X	0.1	= Kilograms-force (kgf; kg)	X	9.81	= Newtons (N)

Pressure

Pounds-force per square inch (psi; lbf/in²; lb/in²)	X	0.070	= Kilograms-force per square centimetre (kgf/cm²; kg/cm²)	X	14.223	= Pounds-force per square inch (psi; lbf/in²; lb/in²)
Pounds-force per square inch (psi; lbf/in²; lb/in²)	X	0.068	= Atmospheres (atm)	X	14.696	= Pounds-force per square inch (psi; lbf/in²; lb/in²)
Pounds-force per square inch (psi; lbf/in²; lb/in²)	X	0.069	= Bars	X	14.5	= Pounds-force per square inch (psi; lbf/in²; lb/in²)
Pounds-force per square inch (psi; lbf/in²; lb/in²)	X	6.895	= Kilopascals (kPa)	X	0.145	= Pounds-force per square inch (psi; lbf/in²; lb/in²)
Kilopascals (kPa)	X	0.01	= Kilograms-force per square centimetre (kgf/cm²; kg/cm²)	X	98.1	= Kilopascals (kPa)
Millibar (mbar)	X	100	= Pascals (Pa)	X	0.01	= Millibar (mbar)
Millibar (mbar)	X	0.0145	= Pounds-force per square inch (psi; lbf/in²; lb/in²)	X	68.947	= Millibar (mbar)
Millibar (mbar)	X	0.75	= Millimetres of mercury (mmHg)	X	1.333	= Millibar (mbar)
Millibar (mbar)	X	0.401	= Inches of water (inH₂O)	X	2.491	= Millibar (mbar)
Millimetres of mercury (mmHg)	X	0.535	= Inches of water (inH₂O)	X	1.868	= Millimetres of mercury (mmHg)
Inches of water (inH₂O)	X	0.036	= Pounds-force per square inch (psi; lbf/in²; lb/in²)	X	27.68	= Inches of water (inH₂O)

Torque (moment of force)

Pounds-force inches (lbf in; lb in)	X	1.152	= Kilograms-force centimetre (kgf cm; kg cm)	X	0.868	= Pounds-force inches (lbf in; lb in)
Pounds-force inches (lbf in; lb in)	X	0.113	= Newton metres (Nm)	X	8.85	= Pounds-force inches (lbf in; lb in)
Pounds-force inches (lbf in; lb in)	X	0.083	= Pounds-force feet (lbf ft; lb ft)	X	12	= Pounds-force inches (lbf in; lb in)
Pounds-force feet (lbf ft; lb ft)	X	0.138	= Kilograms-force metres (kgf m; kg m)	X	7.233	= Pounds-force feet (lbf ft; lb ft)
Pounds-force feet (lbf ft; lb ft)	X	1.356	= Newton metres (Nm)	X	0.738	= Pounds-force feet (lbf ft; lb ft)
Newton metres (Nm)	X	0.102	= Kilograms-force metres (kgf m; kg m)	X	9.804	= Newton metres (Nm)

Power

Horsepower (hp)	X	745.7	= Watts (W)	X	0.0013	= Horsepower (hp)

Velocity (speed)

Miles per hour (miles/hr; mph)	X	1.609	= Kilometres per hour (km/hr; kph)	X	0.621	= Miles per hour (miles/hr; mph)

Fuel consumption*

Miles per gallon, Imperial (mpg)	X	0.354	= Kilometres per litre (km/l)	X	2.825	= Miles per gallon, Imperial (mpg)
Miles per gallon, US (mpg)	X	0.425	= Kilometres per litre (km/l)	X	2.352	= Miles per gallon, US (mpg)

Temperature

Degrees Fahrenheit = (°C x 1.8) + 32

Degrees Celsius (Degrees Centigrade; °C) = (°F - 32) x 0.56

*It is common practice to convert from miles per gallon (mpg) to litres/100 kilometres (l/100km), where mpg (Imperial) x l/100 km = 282 and mpg (US) x l/100 km = 235

Index